LANGUAGES AND ARCHITECTURES
FOR IMAGE PROCESSING

LANGUAGES AND ARCHITECTURES FOR IMAGE PROCESSING

Edited by

M. J. B. DUFF

*Department of Physics and Astronomy,
University College, London, England*

and

S. LEVIALDI

*Consiglio Nazionale delle Ricerche,
Istituto di Cibernetica, Arco Felice,
Naples, Italy*

1981

ACADEMIC PRESS
A Subsidiary of Harcourt Brace Jovanovich, Publishers
London   New York   Toronto   Sydney   San Francisco

ACADEMIC PRESS INC. (LONDON) LTD
24/28 Oval Road, London NW1 7DX

United States Edition published by
ACADEMIC PRESS INC.
111 Fifth Avenue, New York, New York, 10003

British Library Cataloguing in Publication Data

Languages and architectures for image processing.
1. Image processing — Congresses
I. Duff, M.J.B.   II. Levialdi, S.
621.3819'598   TA1632

ISBN 0–12–223320–4

LCCCN 81–67909

Typeset by Oxford Publishing Services, Oxford
and printed in Great Britain by
St Edmundsbury Press, Bury St Edmunds, Suffolk

Contributors

D.M. BALSTON, *Plessey Electronic Systems Research, Southleigh Park House, Havant, Hants., England*

R. BARRERA, *Computing Systems Department, IIMAS, National University of Mexico, Apartado Postal 20–726, Mexico 20, D.F. 58–54–65, Mexico*

J.L. BASILLE, *C.E.R.F.I.A.—U.P.S., 50A, chemin des maraichers, 31077 Toulouse, France*

W. BLACK, *Image Analysis Group, Nuclear Physics Laboratory, University of Oxford, Keble Road, Oxford OX1 3RH, England*

S. CASTAN, *C.E.R.F.I.A.—U.P.S., 50A, chemin des maraichers, 31077 Toulouse, France*

T. CLEMENT, *Image Analysis Group, Nuclear Physics Laboratory, University of Oxford, Keble Road, Oxford OX1 3RH, England*

R.J. DOUGLASS, *Department of Applied Mathematics and Computer Science, University of Virginia, Charlottesville, Virginia 22901, USA*

T.J. FOUNTAIN, *Department of Physics and Astronomy, University College London, Gower Street, London WC1E 6BT, England*

P. GEMMAR, *Research Institute for Information Processing and Pattern Recognition, Breslauer Strasse 48, D–7500 Karlsruhe 1, West Germany*

F.A. GERRITSEN, *Delft University of Technology, Delft, The Netherlands*

G.H. GRANLUND, *Picture Processing Laboratory, Department of Electrical Engineering, Linköping University, S–581 83 Linköping, Sweden*

B. GUDMUNDSSON, *Picture Processing Laboratory, Department of Electrical Engineering, Linköping University, S–581 83 Linköping, Sweden*

A. GUZMAN, *Computing Systems Department, IIMAS, National University of Mexico, Apartado Postal 20–726, Mexico 20, D.F. 548–54–65, Mexico*

J.F. HARRIS, *Image Analysis Group, Nuclear Physics Laboratory, University of Oxford, Keble Road, Oxford OX1 3RH, England*

D.J. HUNT, *International Computers Limited, Research and Advanced Development Centre, Fairview Road, Stevenage, Herts. SG1 2DX, England*

H. ISCHEN, *Research Institute for Information Processing and Pattern Recognition, Breslauer Strasse 48, D–7500 Karlsruhe 1, West Germany*

A. JINICH, *Computing Systems Department, IIMAS, National University of Mexico, Apartado Postal 20–726, Mexico 20, D.F. 58–54–65, Mexico*

Z. KULPA, *Institute of Biocybernetics and Biomedical Engineering, ul. KRN 55, 00–818 Warsaw, Poland*

C. LANTUEJOUL, *Centre du Morphologie Mathematique, 33 rue Saint-Honorè, 77305 Fontainebleau, France*

J.Y. LATIL, *C.E.R.F.I.A.—U.P.S., 50A, chemin des maraichers, 31077 Toulouse, France*

S. LEVIALDI, *Consiglio Nazionale delle Ricerche, Istituto di Cibernetica, Via Toiano 6, Arco Felice 80072, Naples, Italy*

K. LUETJEN, *Research Institute for Information Processing and Pattern Recognition, Breslauer Strasse 48, D–7500 Karlsruhe 1, West Germany*

A. MAGGIOLO-SCHETTINI, *Gruppo Nazionale di Informatica Matematica, c/o Istituto di Scienze dell'Informazione, Università di Pisa, 56100 Pisa, Italy*

R. MANARA, *Elettronica San Giorgio, ELSAG s.p.a. Genova-Sestri, Italy*

R.D. MONHEMIUS, *Delft University of Technology, The Netherlands, (Currently with the Royal Netherlands Navy, Ministry of Defence, The Hague)*

M. NAPOLI, *Gruppo Nazionale di Informatica Matematica, c/o Istituto di Scienze dell'Informazione, Università di Salerno, 84100 Salerno, Italy*

L. NORTON-WAYNE, *Department of Systems Science, The City University, Northampton Square, London EC1, England*

P.L. PEARSON, *Department of Cytogenetics, Sylvius Laboratories, University of Leiden, Wassenaarseweg 72, 2333 AL Leiden, The Netherlands*

E. PERSOON, *Philips Research Laboratories, Eindhöven, The Netherlands*

J.S. PLOEM, *Department of Histo-cytochemistry, Sylvius Laboratories, University of Leiden, Wassenaarseweg 72, 2333 AL Leiden, The Netherlands*

K. PRESTON, JR, *Department of Electrical Engineering, Carnegie-Mellon University, Pittsburgh, Pennsylvania 15213, USA*

T. RADHAKRISHNAN, *Concordia University, Montreal, Canada*

A.P. REEVES, *School of Electrical Engineering, Purdue University, West Lafayette, Indiana 47907, USA*

H.J. SIEGEL, *School of Electrical Engineering, Purdue University, West Lafayette, Indiana 47907, USA*

L.J. SIEGEL, *School of Electrical Engineering, Purdue University, West Lafayette, Indiana 47907, USA*

L. STRINGA, *Elettronica San Giorgio, ELSAG s.p.a., Genova-Sestri, Italy*

G. TORTORA, *Gruppo Nazionale di Informatica Matematica, c/o Istituto di Scienze dell'Informazione, Università di Salerno, 84100 Salerno, Italy*

G. UCCELA, *Gruppo Nazionale di Informatica Matematica, c/o Istituto di Scienze dell'Informazione, Università di Pisa, 56100 Pisa, Italy*

L. UHR, *Computer Sciences Department, University of Wisconsin, 1210 West Dayton Street, Madison, Wisconsin 53706, USA*

J. VROLIJK, *Department of Cytogenetics, Sylvius Laboratories, University of Leiden, Wassenaarseweg 72, 2333 AL Leiden, The Netherlands*

A. WOOD, *Department of Physics and Astronomy, University College London, Gower Street, London WC1E 6BT*

Preface

Large-scale integrated circuit technology is now making a powerful impact on the design of computers. Whereas a few years ago there was virtually no way in which special purpose computers could be economically built to solve particular problems, or to operate in particular problem areas, now it seems not unreasonable for researchers to propose computer architectures strongly optimized for their own applications. The flexibility and cheapness of commercially obtainable microprocessors have brought within the range of quite modest research budgets the chance of configuring a computer for the job in hand.

Whenever the architecture of a computer is radically different from that of the conventional von Neumann type, the programmability of this new machine is a delicate and crucial feature for obtaining high overall performance. Moreover, the programming language for such a computer should both be independent from it (portable and readable) and efficient on it: two truly conflicting requirements! For these reasons, different approaches have been suggested for handling this problem, generally strongly influenced by the previous experience gained in this field by those putting forward their ideas. There is no general consensus, as may be seen by reading the first part of the book, as to whether it is better to program in a high-level language (such as, for example Fortran) and to call from an image processing library subroutines optimally designed for a given machine; or to use an interpreter like APL or PICASSO in which interactivity and accurate diagnostics may be obtained; or, finally, to define a high-level language having specific control structures for local computations and global parallelism, as well as data structures and data types particularly useful in this field. In short, programmability is indeed a key issue in designing a system that can be used for image processing in an efficient yet flexible way.

It is also questionable whether we are sufficiently skilful to build into compilers enough capability to handle efficiently the vast range of problems which can arise. It seems that efficient programming is still likely to require the programmer to have some knowledge of the computer architecture. Alternatively, this knowledge could be built into the structure of the high-level language.

It was to discuss matters such as these that a series of study workshops was initiated in 1979. The first, convened by M.J.B. Duff, was at Cumberland Lodge, Windsor Great Park, England on 4–8 June 1979, and was entitled "Workshop on High-level Languages for Image Processing". Thirty-three specialists in image processing, language and computer design met together to exchange ideas on how high-level languages could be developed or shaped to meet the requirements of image processing and the new computers designed for this purpose. The British Science Research Council provided a generous funding for this workshop.

In the following year, S. Levialdi convened a second workshop entitled "New Computer Architectures and Image Processing" in which the 31 participants reviewed the current scene in computers designed specially for image processing applications. The workshop was held during 2–5 June 1980 on the island of Ischia in the Bay of Naples, and was funded by the Istituto di Cibernetica, CNR (Arco Felice), the Comitato per la Matematica, CNR (Roma), Elsag (Genova), IBM Italia and V.D.A. (Firenze).

A third meeting in this series, convened by Professor L. Uhr and Professor K. Preston, took place in Madison in May 1981 and concerned itself with applications of non-conventional computers. This volume collects together a representative set of papers recording the views of the participants expressed at the first two meetings.

The first half of the book concerns itself primarily with languages, proceeding from the general to the particular, i.e. languages closely linked with particular machines or tasks, some proposals and implementations of high-level languages including concurrency and finally a comparative review of all these approaches. In the second half, the trend is from machines with few processors to machines with many processors, the linking factor being the increasing parallelism. This second part ends with some general discussion and a proposal for evaluating performance.

Every attempt has been made, with the active encouragement of our publishers, to produce this volume as quickly as possible, so as to present an up-to-date account of the state-of-the-art in the emerging field of special computers for image processing and their interaction with language design. The editors hope that their readers will forgive any errors which have escaped the perhaps less than usually careful proof reading that the need for speed has engendered.

London and Naples   _M.J.B. Duff_
May 1981   _S. Levialdi_

Contents

Chapter One

The Interaction between Hardware, Software and Algorithms

Alan Wood

1. INTRODUCTION

The object of this chapter is to investigate some of the complex interactions between hardware and language structures and their effects on the development of algorithms.

The discussion is restricted on the one hand to hardware above the component or implementation level, and on the other to software functions below the operating system. That is, an attempt has been made as far as possible to keep to those system features which are implementation independent. Most of the examples will be taken from image processing problem areas, but the majority of the conclusions drawn apply to all fields of computation. Also, when comparing machine structures there will be a strong bias towards serial versus parallel architectures, this being a reflection of personal experience rather than a statement of relative merit.

One of the central points in this chapter is that the distinction between hardware and software structures is not a clear one, and thus in any discussion, the separation of function between the two is somewhat arbitrary. However, it will be convenient initially to consider the hardware and software components of a computing system separately and then to show how they combine in terms of image processing. The separation used here will be on an intuitive level and will not be given a firm definition.

2. THE DEVELOPMENT OF ALGORITHMS

Solving problems in image processing, as in all other fields of computing, is primarily to do with the search for, and implementation of, algorithms. The algorithms chosen for a particular purpose are strongly influenced by the machine on which they are to be executed. Consequently, the bulk of this section will consider the effects of hardware and software structures on the choice and development of algorithms. It will be seen, in addition, that the architecture of a computing system also affects the underlying thought processes that go into the algorithms.

2.1 Hardware

Given a particular machine structure, a programmer will usually try to find an algorithm to solve a problem in the most efficient way for that machine. (The "most efficient way" will probably mean a compromise between speed and space considerations for the execution of the algorithm, and may also include more complex constraints, such as speed of development, cost of development and other economic factors. Here we shall mainly be considering execution efficiency.)

2.1.1 Algorithmic choice

The structural description of a hardware system can be made at many levels, and all levels contribute towards the choice of algorithm to a greater or lesser degree. Here, two structural features will be looked at representing the lower and higher levels of machine architecture respectively.

2.1.1.1 Data-unit size

The size (or information capacity) of an addressable unit of data can affect the machine user in quite subtle ways. For instance, the word-length of a conventional serial computer defines the accuracy to which the basic arithmetic operations can be performed. This also applies to special arithmetic processors (e.g. floating-point units) that may be attached to the machine. If the machine's accuracy is sufficient for some problem, then obviously these considerations will not worry the user. However, when an application requires a highly accurate final result, and the precision of the processor is insufficient (due to a small word-length say) then an algorithm that has fewer arithmetic operations may be successful due to a reduction in the cumulative error. Alternatively, a means of simulating a larger word by software will be used at the cost of slowing the algorithm down.

When large-scale structural differences between processors are considered (as opposed to the small-scale word size difference just mentioned) then the effects on the programmer become less subtle and more obvious.

In many applications, picture representation in serial machines requires data-reduction techniques both to conserve space and to use more efficient serial algorithms. Such techniques include chain-coding [1] and its derivatives for lines, and quadtrees for scene description [2]. An array machine, on the other hand, would often use the picture without change since it is able, and constrained, to handle the picture as a unit. There would be in general no gain in storage nor in execution speed from either of the above two approaches, and there could well be penalties in both time and space. This is due to the fact that the data reduction techniques mentioned assume an underlying serial von Neumann computer structure with one-dimensional storage units, which is not true of a typical array machine.

2.1.1.2 Processor organization

At the higher level of machine architecture, some of the most striking examples of influence on algorithm design come from basic processor differences, e.g. serial versus parallel.

Three examples follow which attempt to highlight some of the various factors which go into the development of algorithms for different machine structures. The algorithms presented are not put forward as the "best", but as reasonable solutions to the problems. This is particularly true of the examples for the less conventional architectures since they are relatively new and thus ideas are still in the process of development.

The intention is not to go into the algorithms in great detail here, due mainly to space limitations, but where possible examples have been chosen which appear in the literature.

2.1.1.2.1 Outer edges. Given a binary picture, find the outer perimeter of each object. Inner perimeters, i.e. those of holes in the objects, are to be ignored.

For a conventional serial (von Neumann) processor, one method would be to implement an edge extraction algorithm in conjunction with a routine to apply a "containment" relation to the result.

An array processor with a global propagation feature such as the CLIP series [3] would achieve this result by propagating a signal from the edges of the array through the "background" of the picture, until it reaches an object, at which stage the edge points are labelled and the propagation is prevented from entering the object any further. Thus "inner" edges receive no propagation signal and are not extracted. This would be a single machine-level operation in a CLIP style processor.

Global propagation (simulated) would probably be rejected in serial machines due to the necessity for multiple passes and comparisons over the array.

2.1.1.2.2 Local maxima of distance functions. A method of finding those points in a binary image which represent locally maximum values of the distance of object points from the nearest background points, is of importance in thinning and data compression techniques [4]. Figure 1 shows an example of a distance function with the local maxima encircled.

```
( a )                                   ( b )
. . . . . . . . . . . . . . . . . .     . . . . . . . . . . . . . . . . . .
. . . . . X . . . . . . . . .           . . . . . 1 . . . . . . . . .
. . . . . X . . . . . . . . .           . . . . . 1 . . . . . . . . .
. . . X X X X X X . . . . .              . . . 1 1 1 1 1 1 1 . . . . .
. . . X X X X X X . . . . .              . . . 1 2 2 2 2 2 1 . . . . .
. . . X X X X X X X X X . .              . . . 1 2 3 3 3 2 1 1 1 1 . .
. . . X X X X X X X X X . .              . . . 1 2 3 4 3 2 2 2 2 1 . .
. . X X X X X X X X X X . .              . . 1 1 2 3 4 3 2 2 2 2 1 . .
. . . X X X X X X X X X . .              . . . 1 2 3 3 3 2 1 1 1 1 . .
. . . X X X X X X . . . . .              . . . 1 2 2 2 2 2 1 . . . . .
. . . X X X X X X . . . . .              . . . 1 1 1 1 1 1 1 . . . . .
. . X X X . . . . . . . . .              . . 1 1 1 . . . . . . . . .
. . X X X . . . . . . . . .              . . 1 2 1 . . . . . . . . .
. . X X X . . . . . . . . .              . . 1 1 1 . . . . . . . . .
. . . . . . . . . . . . . . . . . .     . . . . . . . . . . . . . . . . . .
```

Fig. 1. (a) Original binary image. (b) distance function (. = O).

The conventional solution to finding the circled points in Fig. 1 would be to calculate the distance function as shown, and then to apply a window operator to the result to find the local maxima. Not allowing for any clever optimizations, this would require of the order of dn^2 operations (where n is the array dimension, and d is the maximum distance) plus at least another n^2 operation for the local maximum function.

For a binary array processor there is a wish to keep to binary operations as far as possible, since arithmetic tends to be relatively slow due to its bit-wise nature. This can be achieved, utilizing the parallel nature of the processor (and of the problem), by a sequence of shrink/expand (erode/dilate) operations [5]. The shrink function, as defined here, is simply the stripping of all object points that have neighbours (8-connected in this case) in the background. The expand function is the addition to an object of those background points with neighbours in the object. The fundamental property of these two functions of importance here is that they are not inverses of each other: a shrink followed by an expand does not in general produce the original picture (Fig. 2). The difference points between a picture and the result of a shrink–expand process upon it, are a subset of the points of the local maximum of the distance function. The process is then repeated for each "level" of shrinking until none of the object remains. Those pixels extracted as difference points at each iteration form the set of local maxima.

(a)   (b)

```
( a )                                              ( b )
. . . . . . . . . . . . . . . . .        . . . . . . . . . . . . . . . . .
. . . . . . . . . . . . . . . . .        . . . . . . . . . . . . . . . . .
. . . . . . . . . . . . . . . . .        . . . . . . . . . . . . . . . . .
. . . . . . . . . . . . . . . . .        . . . X X X X X X X . . . . . . .
. . . . X X X X X . . . . . . . .        . . . X X X X X X X . . . . . . .
. . . . X X X X X . . . . . . . .        . . . X X X X X X X X X X X . . .
. . . . X X X X X X X X . . . . .        . . . X X X X X X X X X X X . . .
. . . . X X X X X X X X . . . . .        . . . X X X X X X X X X X X . . .
. . . . X X X X X . . . . . . . .        . . . X X X X X X X X X X X . . .
. . . . X X X X X . . . . . . . .        . . . X X X X X X X . . . . . . .
. . . . . . . . . . . . . . . . .        . . . X X X X X X X . . . . . . .
. . . X . . . . . . . . . . . . .        . . X X X . . . . . . . . . . . .
. . . . . . . . . . . . . . . . .        . . X X X . . . . . . . . . . . .
. . . . . . . . . . . . . . . . .        . . X X X . . . . . . . . . . . .
. . . . . . . . . . . . . . . . .        . . . . . . . . . . . . . . . . .
```

Fig. 2. (a) The result of shrinking Fig. 1a. (b) The expansion of Fig. 2a.

This technique gives only the location of the maxima and not their values, but a simple parallel incrementation of the distance function at each iteration of the process will supply these values. The advantage of this method for a bit processor is that no arithmetic comparison operations need be done: only logical comparisons (exclusive-OR) and local parallel functions are required (with incrementation if values are important).

For a serial processor, this technique would be much less efficient due to the number of passes over the array required by the successive shrink–expand operations.

2.1.1.2.3 Image rotation. The previous two examples have shown a relatively clear-cut distinction between the algorithms chosen for serial and parallel processors as a solution to a given problem. They have also shown how a different processor structure can affect the choice in a positive fashion. In the case of rotating an image, there seems to be no obvious algorithm for an array processor that is more efficient than the serial technique.

The standard method is to rotate an image by a re-mapping process involving the calculation of each pixel's new co-ordinates followed by a re-assignment of the original pixel's "energy" amongst the elements of the digitization grid. The co-ordinate mapping can be performed using a simple transformation matrix.

For an iterative array processor, the co-ordinates of each pixel are not generally available, although they can be calculated and placed in each pixel's location without too much trouble. The transformation calculation can then be done in a short sequence of parallel arithmetic operations over the whole array. However, the result left in each location is a pair of co-ordinate values (in general, non-integer). There is no obvious parallel method for the subsequent re-partitioning of the result involving, as it does, moving each pixel a non-integral distance and re-assigning the mapped pixels' energies.

The basic reason for this difficulty is that the method of co-ordinate transformation involves operations on the addresses of values rather than on the values themselves. The iterative array processor structure is not designed for this type of operation.

Because of these problems, the user of an array machine is likely to reject immediately, whether consciously or otherwise, problem solutions that require generalized rotation. Thus we also see the effect of new processor structures having a strong influence on algorithm choice in a negative way.

2.1.2 Thought processes

An interesting consequence of the way in which the new architectures give rise to types of algorithm is that new modes of thinking are used in developing solutions. There is more than a mere semantic distinction between the choice of an algorithm and the choice of a method of solution. In the first case, the method chosen is directly influenced by the structure of the machine on which the algorithm is to be executed, whereas in the second case, the solution is developed with greater emphasis on the structure of the problem. Two examples hopefully will clarify this.

2.1.2.1 "Fast" algorithms

Serial computer architectures have given rise to a group of so-called "fast" algorithms as a consequence of their basic functional structure. These algorithms tend to be based on the "divide and conquer" technique, whereby a fast method of calculating a special case of the problem is discovered, say a two-operand case, which is then extrapolated from that to versions of the problem containing multiples of the special case. Probably the best known of these is the Fast Fourier Transform (FFT) [6] in which a simple way of calculating a two-element transform is developed, along with a method for constructing from this solutions for any transform with a power of two number of elements. Other examples include the set of fast matrix multiplication schemes [7].

The implicit assumption for these algorithms is that the cost in time required to bring operands together in order to perform a function is independent of their locations in the computer's data store. This is of course true for von Neumann-type processors, but for an array machine, the time taken to bring the operands to the processing point is proportional to their distance apart in the array. Thus the "fast" algorithms are not generally so fast for an array processor. For instance, the serial time to do a matrix multiply function is proportional to n^3 (or $O(n^3)$), where n is the linear dimension of the matrices for the basic "slow" method. The Strassen algorithm [8] decreases this time complexity to $O(n^{2.81})$. For a CLIP machine, the time for the "slow" method is proportional to n (complexity = $O(n)$) up to the CLIP array size [9]. Similarly, the Fourier Transform in its basic form is an $O(n^2)$ operation. The FFT brings this down to $O(n.\log n)$. Implementing either the "basic" method or the FFT on CLIP results in a linear time complexity, $O(n)$, for both, with a proportionality constant to distinguish between them [10]. (However, the FFT is still a better choice for a CLIP-style machine, since it involves fewer operations per element thus giving a more accurate result.)

The significance of this example is to show that the divide and conquer idea has arisen as a consequence of a processor structure. It is interesting to speculate as to whether this concept would have been discovered if the von Neumann computer structure had not been the prevalent one in computing history, and also whether a more parallel solution to the Fourier Transform will eventually be found due to an interest in parallel machines.

2.1.2.2 "Parallel" thinking

One of the many reasons for the interest in new machine architectures is that if a problem implies a structure for a solution, then it will be solved most effectively on a machine that reflects that structure. Specifically, this is true for parallel processor architectures on which solutions to problems essentially parallel in nature will be executed extremely efficiently. This much is fairly obvious and was realized before any of the parallel machines were actually implemented. In addition, it was discovered that many problems regarded as being serial could be reformulated in a "parallel" way which implied processor types (e.g. operations in Graph Theory [11]). What was not expected before the physical machines appeared was that some parallel solutions to parallel problems for which serial solutions had been developed, actually ran faster on a conventional serial computer which was simulating the parallel machine (via a software emulator), than the serial algorithm running directly on the same machine [12].

The reason why many parallel problems have not been solved with parallel algorithms in the past is that a serial machine (or any other machine structure) imposes its architecture so strongly on users that they are not aware of the parallel solutions. The influence of the von Neumann structure is such that even with an actual, physical, parallel processor for which programs are being written, it requires great effort not to use serial programming techniques in situations where parallel methods are appropriate. Backus [13] has shown how this influence is felt in languages and programming techniques for serial machines.

Thus it can be seen that different architectural concepts affect not only algorithms that run on different machines, but also the way in which the solutions to problems are approached. This is true even if only a conventional processor is available.

2.2 Languages

Experience gained over the years has clearly shown that various problem areas require particular structures in programming languages. This is particularly true of the high-level languages, which in part have been defined by the needs of their fields of use. A few examples are given in Table 1.

Table 1
High-level languages defined by the needs of their fields of use

Field	Functions	Languages
Artificial intelligence	List processing	LISP
Business	File handling	COBOL
Scientific	Numeric	FORTRAN, ALGOL etc.

It is wrong though to consider languages as being separate from hardware in concept. In fact, languages are software machines because in principle any function of hardware can be implemented in software and vice versa. One need only consider how languages are often used as means for extending hardware features, to see how this principle is used, however tacitly.

Thus, language structures have effects on algorithms which are virtually identical to those listed above in connection with hardware. Most of that which has been stated with regards to hardware and algorithms applies to languages and algorithms, and so will not be restated here. The next section deals with some of the implications of this.

2.3 Image processing systems

The emergence of the new processors from image processing research stems from the failure of conventional processors to cope efficiently with the problem areas in this field. Since the input to most image processing systems is an image (often a two-dimensional digitized array), the ability to handle images, however defined, as opposed to a collection of pixels seems to be a fundamental requirement. Thus software and hardware structures have been proposed in many places for this, along with sets of operations to act upon the image units.

Therefore, image processing has tended to define both hardware and software structures; i.e. it suggests *system architectures*. Consequently, the job of the system designers is to specify the hardware and software components of the basic system and, most importantly, the interface between them.

When mis-matches exist between the hardware and software substructures, many difficulties occur in the overall system. This can be seen very clearly in the situations where serial and parallel languages and machines are mixed. For instance, imposing a serial programming language on a parallel processor is generally unsuccessful unless the language contains, in addition to its serial constructs, only low-level parallel operations or sub-routine calls. In this case, the result is effectively a serial system with a special purpose addition for accomplishing some tasks efficiently, and the new processor structure is not fully utilized.

On the other hand, if a language suited to parallel concepts is implemented on a serial machine, the run-time efficiency drops considerably due to the time required for interpretation. This can be partially overcome by replacing the interpreter with a compiler, but then the degree of mis-match will be felt most strongly by the compiler writer as he attempts to bridge the architectural gap. In some cases this gap becomes so wide that a reliable and efficient compiler cannot be produced.

Of course no machine, language or anything else is totally serial or totally parallel: serial computers have parallel arithmetic units, and parallel computers act on sequences of instructions. But the difficulties encountered in trying to define languages for the new machines arise mainly from the distance between the well-known languages and the new architectures, on the serial–parallel scale. The inevitable conclusion to be drawn then, is that for the optimum efficiency for compilation and execution in a complete system, machine and languages must be as similar in structure as allowed by the requirements of flexibility.

Two systems described at the workshops seem to fall into the category of matched structures: the Cytocomputer [14], and the LISP data-flow machine [15]. In the case of the Cytocomputer, a language was found in mathematics [16] which fitted the structure of the machine rather well. For the LISP machine, the opposite situation occurred whereby an existing (non-von Neumann) language suggested a (non-von Neumann) computer structure. Outside image processing, similar situations occur with, for instance, reduction languages [17] and recursion [18] suggesting machine architectures.

Perhaps this conclusion can be emphasized with the example of assembly languages. These produce in theory, the most efficient programs at run-time, and assemblers are much easier to write (and usually faster to run) than compilers. This is due to the extremely close relationship between the language and machine architectures. So, why not write everything in assembly language? The problem is that these low-level languages, whilst very efficient to run and assemble, are inefficient tools for program development, especially as programs become moderately large. It has been commented that the rate at which debugged program statements are produced is independent of the language in which they are written, and so high-level languages have an advantage over low-level ones due to the many-to-one relationship between low and high-level statements. But high-level languages often produce code that is less efficient at run-time than the equivalent assembly-level program.

There seems to be then, a vicious circle in which the choices are run-time, or developmental, efficiency. With careful design of languages and systems of optimizing compilers, some of the run-time difficulties can be alleviated, but the fundamental problem remains. This is that the von Neumann machine structure is too low-level for current programming requirements.

The solution then, is to design future machines with higher-level architectures, probably taking concepts from high-level language research. This is, of course, no trivial task, but a degree of success has already been achieved in this as is seen by some of the various machines described at these two workshops.

SUMMARY

The main argument in this chapter is on two levels:

(i) The description of an image processing, or any other, machine architecture must include both hardware and software components to be complete.

Hardware and software are "duals" in that the functions of one can, in theory, be implemented in the other. Thus the division of function between them is more a matter of convenience and economics, than of basic principle.

(ii) While the hardware/software division is being decided upon, the interface between the two must be considered. Just as in the case of interfaces between hardware units, if the two halves are too disparate in function, then the interface becomes unwieldy and difficult to design, thus allowing greater scope for unreliability. In the limit the interface will not work at all.

Therefore, the hardware and software must be at a level where they are well matched. Traditional machines have too low a level of hardware to match currently required levels of software, and so new machine architectures should take into consideration the needs of possible high-level languages which they are to execute. It is certainly true that the most successful languages in the past have been those whose structure and level have most closely reflected the von Neumann machines, e.g. assemblers, FORTRAN, BASIC etc.

Image processing is fortunate in that it is a relatively new field with relatively few commercial constraints compared to large computer manufacturers, due to its base in the academic environment. Thus there is at present a greater freedom to experiment with new architectural ideas than exists in the commercial world. If this freedom is not exploited now, then it will be lost when image processing becomes "big business" with its attendant reluctance from users to change.

REFERENCES

[1] Freeman, H. (1974). "Computer processing of line-drawing images". *Comput. Surveys* **6**.
[2] Klinger, A. and Dyer, C. R. (1976). "Experiments in picture representation using regular decompostion". *Comp. Graph. & Image Proc.* **5**, 68.
[3] Duff. M. J. B. (1978). "Review of the CLIP image processing system". Proc. Nat. Comp. Conf. Anaheim, USA, pp. 1055–1060.
[4] Davies, E. R. and Plummer, E. P. N. (1980). "Thinning algorithms and their role in image processing". Proc. BPRA Conf. on Patt. Recog. Oxford (To be published in *Patt. Recog.*).
[5] Wood, A. M. (1980). "The CLIP4 array processor". *J. Brit. Interplanetary Soc.* **33**, 338.
[6] Brigham, E. O. (1974). "*The Fast Fourier Transform*". Prentice-Hall, Englewood Cliffs.
[7] Kronsjo, L. I. (1979). "*Algorithms: their complexity and efficiency*". Wiley-Interscience, New York.
[8] Strassen, V. (1969). "Gaussian elimination is not optimal". *Num. Math.* **13**, 354.
[9] Klete, R. (1978). "Fast matrix multiplication by Boolean RAM in linear storage". *Math. Foundations of Comp. Sci.* **1978**, 308.
[10] Ip, H. H-S. (1980). "An analysis of the two-dimensional Fourier Transform for a typical parallel processor". Int. Proj. Rep. Dept. of Physics and Astronomy, University College London.
[11] Levitt, K. N. and Kautz, W. H. (1972). "Cellular arrays for the solution of graph problems". *Comm. ACM* **15**, 789.
[12] Michie, D. (1980). "Expert systems". *Computer J.* **23**, 369.
[13] Backus, J. (1978). "Can programming be liberated from the von Neumann style? A functional style and its algebra of programs". *Comm. ACM* **21**, 613.
[14] Lougheed, R. M. *et al.* (1980). "Cytocomputer architectures for parallel image processing". Proc. IEEE Workshop on Picture Data Description and Management, Asilomar, USA, pp. 281–286.
[15] Guzmán, A. and Segovia, R. (1976). "A parallel configurable LISP machine". *Tech. Rep.* **133**(7), Comp. Sci. Dept., University of Mexico.
[16] Matheron, G. (1975). "*Random sets and integral geometry*". Wiley, New York.
[17] Mago, G. A. (1979). "A network of microprocessors to execute reduction languages". *Int. J. Comp. & Inf. Sci.* **8**, Part I, p. 349; Part II, p. 435.
[18] Glushkov, V. M. *et al.* (1974). "Recursive machines & computer technology". Proc. IFIP Cong., Stockholm.

Chapter Two

PICASSO, PICASSO–SHOW and PAL —A Development of a High-level Software System for Image Processing

Z. Kulpa

1. GENERAL DESIGN PROBLEMS

Generally speaking, to construct a programming language or system one should take into consideration the following problems:

(i) *Data structures:* types of data structures, their description, definition and machine representation;

(ii) *Data manipulation:* basic operations possible on required data structures, their description, definition and methods for their machine realization;

(iii) *Program structure:* basic control-flow structures, program structuring mechanisms, their textual form, definition and machine realization.

Before attempting to solve these problems for some specialized language, one should analyse the field of its application, i.e. state what kind of processing the language should provide. This is shown schematically in Fig. 1 for the case of image processing. More specific characterization of the four "processing boxes" of the figure is as follows.

1.1 Picture processing

(i) Mostly *local parallel* operations
(ii) Homogenous processing of all pixels
(iii) Use of special non-computer devices (e.g. optical processing)

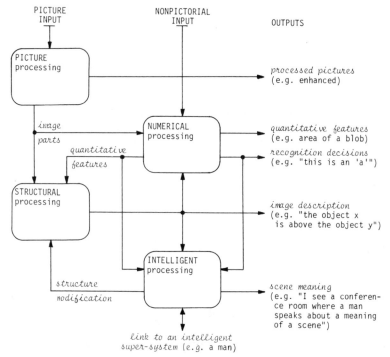

Fig. 1. Schematic structure of an image processing system considered as a subsystem of an intelligent machine. Only main information-flow links are shown.

(iv) Pictorial input/output problems
(v) The main problems in computer processing include:
    (a) handling of huge amounts of homogenous data with inherently *two-dimensional* structure;
    (b) effective packing of information in memory;
    (c) effective implementation of parallel operations.

1.2 Numerical processing

  (i) Arithmetic;
 (ii) Simple logical (Boolean) operations;
(iii) Comparatively small quantities of data, practically without structure (numerical and logical variables);
(iv) No specific problems with operation repertoire;
 (v) Covered excellently by ordinary "general purpose" languages (e.g. ALGOL, PASCAL).

1.3 Structural processing

   (i) Highly entangled data structures (lists);
  (ii) Rather large amounts of data, mostly of structural character (references
       or links to data items);
 (iii) Basic operations: access to structure elements (link following);
 (iv) Problem of choice of basic list element and its computer representation;
  (v) Sophisticated memory management system needed.

1.4 Intelligent processing

   (i) Data interpretation based on stored knowledge (world model);
  (ii) Knowledge-based inference and reasoning;
 (iii) Various artificial intelligence devices;
 (iv) Based mainly on structural processing.

Moreover, three general requirements on any reasonably high-level image processing language should be considered also:
   (i) *Flexibility:* e.g. any number and types and dimensions of pictures—eventual restrictions can be introduced in some particular implementation, not in the language itself!
  (ii) *Efficiency:* parallel image processing; transparency to the machine level; effectively translatable language constructs.
 (iii) *Universality:* all classes of processing covered; eventually more specialized sub-languages can be defined.

Taking into account all the above, the language design problems can be framed as follows:
   (i) *Data structures:* numerical types; pictures; list structures; static (declared) versus dynamic (generated) data items; very effective machine representation, especially of pictures.
  (ii) *Data manipulation:* standard arithmetic; operations of access to data structure elements; specific picture operations (e.g. local parallel); very effective machine realization, especially of picture operations; extendability of operations repertoire.
 (iii) *Program structure:* basic control mechanisms: serial execution, choice statement, loop statement, subroutine (procedure) constructs; introduction of parallelism: "parallel for" statement, colateral execution mode, co-routines; interactiveness of the language.

The above issues are illustrated on an example of the development of a high-level software for the universal image processing system CPO-2/K-202 [1–3]. The work was concentrated around three main subjects:
   (i) PICASSO (PICture ASSembly-programmed Operations)—the large

library of subroutines, realizing basic operations on pictures, and still growing in size [1–3];

(ii) PICASSO–SHOW—a family of medium-level interactive languages (versions no. 1, 1.5 and 1.6 (presently in use), 2 (not implemented) and 3 (implementation in its final stages)) [2–7];

(iii) PAL (Picture Analysing Language)—a high-level language, PL/1-like, semi-formally defined, some parts implemented experimentally [7–9].

The above subjects are heavily interdependent, e.g. the PICASSO library constitutes part of the PICASSO–SHOW as well as PAL systems; the dynamic memory allocations system implemented for PAL is now being included into PICASSO–SHOW interpreter, etc.

2. THE PICASSO SUBROUTINES LIBRARY

The library is written in assembly code for the K-202 Polish minicomputer and currently comprises about 170 operations [1–3]. All operations assume the same structure of processed data items—numbers, pictures and number vectors. They are written so as to achieve maximum efficiency in execution time. The program listings and internal structure are partially standardized in order to become self-documented and allowing their easy inclusion into the PICASSO-SHOW interpreter (see p. 18).

Basic data structures, namely pictures, are rectangular matrices of pixels, and to achieve greatest flexibility, they can have any dimensions and any number of bits of pixel values representation. In the memory, every picture is preceded by a header including the following parameters:

X0, Y0:   co-ordinates of the lower left corner of the picture (in some absolute co-ordinate system);

M, N: width and height of the picture (in pixels);

S: the number of bits per pixel;

L: the length of picture representation (in memory words).

Every picture operation uses this header to organize appropriately its processing of the picture.

Two different representations of pictures in memory are used: so-called "packed" and "stacked" representations.

For the packed representation, all S bits representing the pixel value are stored in S consecutive bits of the same memory cell; one such cell contains several pixels usually.

For the stacked representation, the picture is stored as S binary "planes", each containing a single bit of the representation of all pixels of the picture. Every memory cell in the plane contains the given bit of W consecutive pixels (along a row) of the picture, where W is the machine word length. The S bits of representation of some pixels are stored in S different memory cells, placed in the same positions of different planes. A binary picture (S = 1) is a special case of a stacked picture, and consists of a single plane.

The whole input picture stored in the picture buffer memory of the CPO-2 device is a stacked picture with parameters X0 = Y0 = 1, M = N = 512, S = 4 (i.e. 16 grey-levels) and L = 64 K words.

Utilizing the above packing of pixels into words and the fact that computers usually perform most operations with a single instruction over the whole word (bit-parallel), many PICASSO subroutines implement a semi-parallel processing method, gaining significantly in speed and efficiency over more serial processing requiring individual access to every single pixel.

Most PICASSO picture operations are written in two versions, one for packed and one for stacked arguments. Some of them have simplified versions operating on binary pictures. There are also operations processing only binary pictures (e.g. many propagation operations). The whole library is actually divided into 14 groups:

- (i) CPO-2 input/output (e.g. SCAN: scanning a picture window from the image memory);
- (ii) changing the form of pictures (e.g. TOPACK: change from stacked to packed representation);
- (iii) single-pixel (serial) operations (e.g. SREAD: read a pixel value; BAPROX: polygonal approximation of a binary contour);
- (iv) input/output (e.g. SDUMPV: print visually a stacked picture; SLOADN: load numerical form of a stacked picture from paper tape);
- (v) global picture parameters calculation (e.g. SHIST: calculate a grey level histogram; BAREA: calculate an area of a blob);
- (vi) one-argument picture operations (e.g. SNEG: negate a picture);
- (vii) two-argument picture operations (e.g. SAND: and-ing two stacked pictures);
- (viii) shifting operations (e.g. SSH1X: shift a picture by one point);
- (ix) testing operations (e.g. SEQ: are two pictures equal?);
- (x) local (parallel) operations (e.g. BCONT8: extract a 4-connected contour);
- (xi) propagation operations (e.g. BCON4: extract required 4-connected component);
- (xii) object-extraction operations (e.g. LOCTHR: determine local thresholds for dynamical thresholding);
- (xiii) generation operations (e.g. SCIRC: generate a circular disk on a picture);
- (xiv) picture correction operations (e.g. CORSHW: additive shading correction).

3. THE PICASSO–SHOW INTERACTIVE LANGUAGES

General aims of the PICASSO–SHOW languages construction were, at the beginning, to make a language which:

  (i) has a simple structure and a small interpreter;
 (ii) is easily extendable with new operations;
(iii) can use, in simple manner, the PICASSO library as a set of basic operations;
 (iv) assures interactive execution of individual operations from its repertoire;
  (v) can serve as a tool for convenient composition of medium-complexity picture processing programs;
 (vi) assures also non-interactive execution of such composed programs;
(vii) is easily translatable manually into assembly language (so as it can serve as a convenient tool for development of new PICASSO sub-routines).

The first operational version of the language was made in 1976 [see 1, 4], and then improved in the summer of 1977, resulting in the versions named 1.5 and 1.6, exploited till now. At the same time, the works on its extension were continued, producing the projects No. 2 and 3, the latter being now in the final stages of implementation.

The PICASSO–SHOW 1.6 language is characterized by the following features [see 1, 3, 4 for more details]:

  (i) Two execution modes: *interactive* (immediate execution of every single statement printed) and *interpretative* (sequential execution of a list of statements, i.e. a program, stored in memory).
 (ii) Data types:
      (a) *stacked* (S0, ..., S9) and *packed* (P0, ..., P9) *pictures*, of any sizes and numbers of bits per pixel;
      (b) *vectors of integers* (R0, ..., R9);
      (c) *integer variables: simple* (V0, ..., V9) and *vector elements* (R0(i), ..., R9(i)).
(iii) Dynamic declarations (generators) of pictures and vectors.
 (iv) Predefined set of standard names of pictures, vectors, variables and labels.
  (v) Uniform structure of all statements (name + list of parameters separated by commas).
 (vi) Easy extendability of the statement repertoire by adding a subroutine to the interpreter library and standardized description of its parameters into interpreter "operation vocabulary".
(vii) Program flow of control organized by jump statements, loop statements and sub-program call/return statements.

(viii) No arithmetic expressions (single assembler-like language statements for every arithmetic operation).

(ix) Real arithmetic realized by a set of operations acting on triplets of integer variables.

The third version of the language [3, 5–7] is substantially extended and changed. Main extensions are listed below:

(i) Three (instead of two) execution modes: *interactive* (as before), *interpretative* (as before, but with full editing possibilities on the stored program) and *program* or *fast-interpretative* (a program is translated firstly to an intermediate code, whose interpretation is much faster).

(ii) Additional data types:
    (a) *long integers* (L0, ..., L9, VL0[i], ..., VL9[i]),
    (b) *reals* (R0, ..., R9, VR0[i], ..., VR9[i]),
    (c) separate *binary pictures* (B0, ..., B9),
    (d) *atoms* (A0, ..., A9, W0[i], ..., W9[i]),
    (e) *atom vectors* (W0, ..., W9).

(iii) Arithmetical expressions can be used as numerical parameters.

(iv) Global and local labels in programs; global labels can have any mnemonic names.

(v) List processing facilities [6]:
    (a) *dynamically extendable* (and contractable) *vectors*,
    (b) *atoms as list elements:* an atom is a dynamically changeable set of "fields" (i.e. a dynamic record); every field has a name (field selector) and a value; the field values can be of any type (determined by the field selector); using atoms (strictly: references to atoms) as field values any list structures can be built up,
    (c) *atom vectors* (i.e. vectors with atoms as elements).

(vi) Object generators can be used as parameters—it produces unnamed objects of local use.

(vii) Accessible loop-controlled variables (K0, ..., K9).

(viii) Possible access to picture and vector header parameters (read-only).

(ix) Integration of operation vocabulary entries and subroutine body into single "operation modules".

(x) Full dynamic storage allocation system, implemented previously for the PAL compiler [7].

A simple example of a program in PICASSO–SHOW 1.6 language follows. Texts in square brackets are comments.

```
[8-CONNECTED CONTOUR EXTRACTION FROM BINARY
PICTURE S0 ONTO S2]
$0: [A LABEL]
  SPUT,S2,0,        [S2 := 0]
```

```
SET,V1,1,              [V1 := 1]
BEGLOOP1,4,            [4-TURN LOOP]
  SSH1D,S0,S1,V1,     [SHIFT OF S0 IN DIRECTION V1]
  BDIF,S0,S1,S1,      [S1 := S0 & ( ¬S1)]
  BOR,S2,S1,S2,       [S2 := S2 ∨ S1]
  ADD,V1,2,           [V1 := V1 + 2: NEW DIRECTION]
ENDLOOP1,
DISP,S2,1,1,4,        [DISPLAY THE RESULT ON THE SCREEN]
DO,3,                 [RETURN TO INTERACTIVE MODE]
##
```

In PICASSO–SHOW 3 notation it would look like the program below (most comments are omitted here):

```
$CONT4: SPUT,B2,0,
        BEGLOOP1,1,7,2, <LOOP FROM 1 UNTIL 7 WITH STEP 2>
        SSH1D,B0,B1,K1,
        BDIF,B0,B1,B1,
        BOR,B2,B1,B2,
        ENDLOOP1,
        DISP,B2,1,1,4,
        DO,3,
##
```

Instead of this program a single instruction, calling the PICASSO operation BCONT4 will also do the job:

In 1.6 version: BCONT4,S0,S1,C4(1,1),

In 3rd version: BCONT4,B0,*B0,B2 ↑ 4[1,1],

4. THE HIGH-LEVEL LANGUAGE PAL

The PAL language [8, 9] is a rather complex high-level programming language for picture analysis. Generally, it was intended that the language should:

(i)   Be high-level (somewhere near ALGOL 68), but
(ii) Allow effective execution of complex picture processing programs, through:
      (a) effective representation of basic data structures, especially picutres;
      (b) transparency to the machine level (machine-oriented constructs besides, but not replacing, the high-level ones);
      (c) possibility to specify parallel picture processing.
(iii) Be simply extendable with new operations, eventually grouped into integrated software packages (application or subfield modules).
(iv) Have list-processing facilities to process structural descriptions of complex pictures (e.g. natural scenes).

(v) Be structured so as to allow comparatively simple working out of its interactive version.

(vi) Assure safe execution (i.e. it should have no undetectable error conditions) as well as allowing an introduction of user-defined error conditions in a standardized manner.

(vii) Have precise, mostly formalized but readable definition.

On the basis of a broad analysis of data structures, operations in the image-processing field, other languages proposed earlier (PICTURE ALGOL [10], PAX [11], SOL [12]), and general design rules for programming languages, a proposal of such a language was created. Although it was (almost) completely and semi-formally defined [8, 9], it should not be considered as a fully finished result, but rather as some elaborate exercise in creating such a language.

In effect, the PAL is a high-level language, although allowing considerable control of a programmer over the machine execution of a program. The high-level constructs include, among others:

(i) block structure, although more flexible than in ALGOL-like languages;

(ii) "case clauses", allowing to choose between many execution alternatives (selected by integer value);

(iii) "iterative clauses", of two sorts:

(a) classical arithmetic progression "for statements", very similar to an ALGOL 68 construct,

(b) "for all" statements, allowing execution of a given statement for all elements of a vector or all points of a picture, also in parallel (if specified);

(iv) procedure (also co-routine) and operator definition mechanisms (somewhat similar as in ALGOL 68), allowing easy extension of language operations repertoire;

(v) expressions, not only artithmetical, but more general "formulas" including diverse operators, standard or user-defined (like in ALGOL 68),

(vi) module mechanism, allowing grouping interdependent operators and procedure definitions as well as data declarations into a single "library module", and extending in such a way the standard operation set of the compiler.

Data types (called here, as in ALGOL 68, *modes*) of the language include:

(i) *arithmetical modes* (INTeger and REAL, with various precisions (lengths));

(ii) *complex modes:*

(a) dynamic vectors (*rows*: (σ)ROW, where σ is any simple mode);

(b) *pictures*, of various kinds: BINary, PACKed, STACKed (see

p. 16) and also with INT and REAL elements (pixels);

(c) *atoms*, very similar to that in PICASSO–SHOW 3 (see p. 19), serving as list structures elements;

(iii) other *simple modes*:

(a) *references* to the objects (REF μ, where μ is any mode), indispensable to built list structures;

(b) *pairs* of integers (PAIR), needed to specify neighbourhoods and other items in local picture operations;

(c) *characters* (CHAR), used in alphanumerical input/output;

(d) *co-routine* (COROUT), needed in manipulation with co-routines;

(e) machine-oriented modes: WORD (a *machine word*) and ADDRESS (physical *address* of a memory cell).

There are various language constructs allowing access to and manipulation with the elements of data structures: *slices* allow extraction of single picture points, row elements as well as any subpictures (windows) or row fragments; *selections* allow access to atom fields and alteration of their composition; *variations* allow dynamic change of row length; *assignations* allow a change of value of the whole structure (variable of some mode). *Denotations* provide means for writing down explicitly the value of any required mode. The *objects* (*variables*) of any mode can be *declared* (i.e. generated at compile time), or *generated* at a given moment in program run-time (for dynamic objects, e.g. list structure elements). Object declarations and generators can be placed any-where in the program, not only at the beginning of a block—it facilitates the construction of an interactive version of a language.

Transparency to the machine level is obtained by including machine-oriented modes (WORD and ADDRESS) and allowing the use of machine instructions within the program. Also the programmer has some control over storage allocation mechanisms, by introduction of so-called "object types":

(i) *Constants:* objects with declared and not changeable value (e.g. CONST REAL pi = 3.141592653);

(ii) *Global* and *local variables:* objects with declared and not changeable structure (dimensions), but changeable value; locals and globals differ in their life-time: for locals it is a block, and for globals the whole program;

(iii) *Dynamic variables:* objects with dynamically computable dimensions and values; their life-time is not connected with the program text structure—they can be generated and explicitly (or implicitly, by garbage collector) destroyed in any moment of the run-time.

A special mechanism of run-time error detection and error reaction allows safe execution. New, non-standard error conditions and error reactions can be specified or redefined by the programmer also.

The description of the language is divided into three levels:

(i) *Syntax:* a context-free grammar, written in van Wijngaarden notation.

(ii) *Static semantics:* essentially syntactical specifications, but not definable by a context-free grammar; it is described with a set of formulas in formal logical calculus.

(iii) *Dynamic semantics:* a program meaning, in terms of actions that should be performed by a computer in order to execute the given program; it is described in somewhat formalized natural language (in [8, 9] it is Polish).

As an example, consider the following procedure definition, describing a contour extraction algorithm, the same as in the previous section:

```
REF B PICT PROC bcont4 (REF B PICT b0)
  BEGIN
    REF B PICT b2 := ↑ b0; #The standard " ↑ " operator generates a
                                         new copy of a picture #
    b2 VAL 0;
    FOR ALL dir FROM {[1,0], [0,1], [−1,0], [0,−1]}
      DO b2 ∨ ↑ b0 ＼(b0 SH dir) OD;
    b2
  END #bcont4#;
```

or, using other language possibilities:

```
REF B PICT PROC bcont4 (REF B PICT b0)
  BEGIN
    REF B PICT b2 := ↑ b0;
    PAR FOR ALL p IN b0'
      DO b2' ⊥p := b0' ⊥p & ¬(b0' ⊥p[1,0] & b0' ⊥ p[0,1] &
                                        b0' ⊥p[−1,0] & b0' ⊥p[0,−1])
          ON border ERROR REACT 0
      OD;
    b2
  END #bcont4#;
```

ACKNOWLEDGEMENTS

The research reported here was supported by the Research Programme No. 10.4.

REFERENCES

1] Kulpa, Z., Dernałowicz, J., Nowicki, H. T. *et al.* (1977). "System cyfrowej analizy obrazów CPO-2" (CPO-2 digital pictures analysis system, in Polish). Inst. of Bio-cybernetics and Biomedical Engineering Rep., Vol. 1. Warsaw.

[2] Kulpa, Z. and Dernałowicz, J. (1978). "Digital image analysis system CPO-2/K-202, general hardware and software description". Proc. IVth Polish-Italian Bioengineering Symp. on Patt. Recog. of Biomedical Objects. Ischia, Italy.

[3] Kulpa, Z., Dernałowicz, J., Nowicki, H. T. and Bielik, A. (1981). "CPO-2/K-202—a universal digital image processing and analysis system". In *Digital image processing systems*, Bolc, L. and Kulpa, Z. (eds). Lect. Notes in Comput. Sci., Vol. 109, Springer-Verlag, Berlin.

[4] Kulpa, Z. and Nowicki, H. T. (1976). "Simple interactive picture processing system PICASSO-SHOW". Proc. 3rd IJCPR, Coronado, USA, pp. 218–223.

[5] Nowicki, H. T. (1978). "Interactive picture processing language PICASSO-SHOW 3 and its interpreter". Proc. IVth Polish-Italian Bioengineering Symp. on Patt. Recog. of Biomedical Objects. Ischia, Italy.

[6] Kulpa, Z. (1978). "Propozycja podjęzyka przetwarzania list do systemu PICASSO-SHOW" (A proposal of a list-processing sublanguage for the PICASSO-SHOW system, in Polish). Inst. of Biocybernetics and Biomedical Engineering Int. Rep., Warsaw.

[7] Bielik, A. and Kulpa, Z. (1978). "System dynamicznej rezerwacji pamięci i przetwarzania listowego SETSYS/K-202" (Dynamic storage allocation and list processing system SETSYS/K-202, in Polish). Inst. of Biocybernetics and Biomedical Engineering Int. Rep., Warsaw.

[8] Kulpa, Z. (1977). "Język analizy obrazów graficznych PAL" (A graphic pictures analysing language PAL, in Polish). Inst. of Biocybernetics and Biomedical Engineering Int. Rep., Warsaw.

[9] Kulpa, Z. (1979). Konstrukcja języka programowania algorytmów cyfrowego przetwarzania złożonych obrazów wizualnych" (Design of a programming language for digital processing algorithms of complex visual images, in Polish). Ph.D. Thesis, Institute of Foundations of Computer Science, Warsaw.

[10] Kulpa, Z. (1972). "A picture processing system PICTURE ALGOL 1204". Proc. 7th Yugoslav Int. Symp. on Inf. Proc. (FCIP'72), Bled.

[11] Johnston, E. G. (1970). "The PAX II picture processing system". In *Picture processing and psychopictorics*, Lipkin, B. S. and Rosenfeld, A. (eds). Academic Press, New York and London.

[12] Schwebel, J. C. (1972). "A graph structure model for structural inference". In *Graphic languages*, Nake, F. and Rosenfeld, A. (eds). North-Holland, Amsterdam and London.

Chapter Three

Design of a High-level Language (L) for Image Processing

T. Radhakrishnan, R. Barrera, A. Guzmán and A. Jinich

1. INTRODUCTION

The language L is an extension to ALGOL and it is intended for programming image processing applications. It was developed with the specific problems of multispectral remote sensing data analysis in mind and is a natural outgrowth of the experience obtained in the design and implementation of the PR System [1]; it is introduced with the objective that an individual with the knowledge of both a programming language and the PR problems should be able to solve these problems rather "easily" with L. Furthermore, we hope that such programs in L will be shorter and more "readable" than those in a POL (procedure oriented language). This approach of extending a POL to facilitate programming a specific class of problems is not totally new [2]. In fact, a language for picture processing, PAX [3], already exists. PAX was designed in the 1960s as an extension (a set of subroutines) to FORTRAN. But since no new data type was introduced in PAX, a programmer still has to do a lot of book-keeping. It must be kept in mind that all image handling within L is done as if the whole image exists in random access memory; therefore, it is a problem of implementation to make the transfers to and from the disc that will undoubtedly be required.

In this report, we specify only those parts of L which are different from ALGOL. Knowledge of ALGOL and the use of syntax diagrams are assumed [4, 5].

Nothing particular about ALGOL, other than it is used in the present PR implementation, has guided our base-language selection. Most probably other languages like PASCAL, ALGOL-W, or PL-1, with record and reference handling capabilities, would be more suitable for general image processing applications [6].

By way of extension, the language L includes the following:

(i) Six new data types (four categories)—Images, Image Groups; Boolean Images and Boolean Image Groups; Window; Virtual-Image, Virtual Image Groups.

(ii) A set of additional declarations to define, to store, and to process images conveniently.

(iii) A set of unary and binary operators on images and ways of defining predicates with the new data types.

(iv) The concept of "attribute" is introduced to define an integral property of an image. An attribute will have a value as determined by the associated procedure known to the compiler. Provisions are made in L to access, compute, and to store such attributes. This has been found valuable, from the earlier experiences, in solving PR problems. An important property of attributes is that they provide in-built documentation and concise input and output.

2. DECLARATIONS

In this section, we describe all the additional (to ALGOL) declarations. First, we present the motivation for a declaration, then we follow it with its syntax. An example is included for easy understanding.

In the syntax definitions, the following syntactic units of ALGOL are used (when it is necessary to distinguish among the different occurrences of a syntactic unit, subscripts are used as ID_1, ID_2, etc.):

IC: an integer constant;

ID: an identifier (a simple name);

IV: an integer variable (a name that can take integer value).

(a) "Bits per pixel" is introduced to define the number of bits required to store one pixel of an image. We recommend that it be an integral multiple of the byte length of the computer for ease of implementation:

Syntax:

e.g. BITS PER PIXEL: 8;

(b) An image is a collection of one or more rectangular arrays of pixels. The following declaration is self explanatory.

Syntax:

DEFAULT IMAGE SIZE: < ⟶ IC ⟶, ⟶ IC ⟶ > ; ⟶

e.g. DEFAULT IMAGE SIZE: <1024,520>;

(c) The data type image is declared by the following declaration. While programming in L, a programmer need not treat an image as a collection of pixels, if preferred.
Syntax:
Let the term UT be defined as

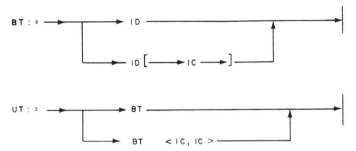

Then image is defined as follows:

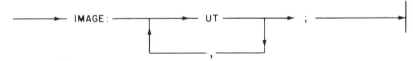

e.g. IMAGE: A,B<100,200>;
     IMAGE: C, D[3], E[5]<10,20>;

Each rectangular array of pixels is called a *band*. An image has *one* or *more* bands. The above example defines the images A and C each with one band and D with 3 bands, all having the default image size. Additionally B is defined with one band of size, 100 × 200 pixels, and E is defined with five bands, each of which has a size of 10 × 20 pixels. It is clear that all bands of a single image have the same rectangular size.

(d) A window is defined to position and view a (rectangular) portion of an image. Some operations on windows and defining its *frame size* are discussed in section 4.
Syntax:

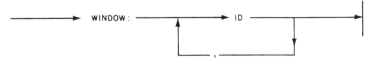

e.g. WINDOW: P,Q,R;
The concept of image group is similar to the one-dimensional arrays in FORTRAN or ALGOL. It is introduced to consider a group of images of identical dimensions and to process them with the use of subscripts. Let

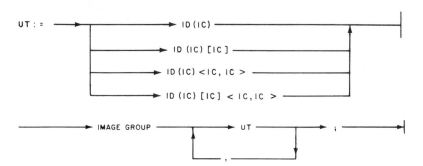

e.g. IMAGE GROUP: A(3), B(2)[3];
      IMAGE GROUP: C(3)<100,200>, D(2)[3]<40,10>;
The image group A has three components and each of them has a single band; the image group B has 2 components and each of them has 3 bands. They all have default image size. The image groups C and D are similar to A and B respectively; but their sizes are different, as shown in the declaration.

(e) As stated in section 1, an image may be associated with one or more attributes. Let Ω denote the set of all attributes known to the compiler of L. Let Attr$\epsilon\Omega$ and

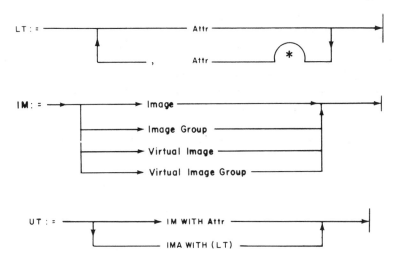

ASSOCIATE ATTRIBUTE: → UT → ;

e.g. ASSOCIATE ATTRIBUTE: A WITH HISTO,
                          B WITH (HISTO,VAR),
                          C WITH VAR;

(f) Virtual image is a concept quite useful in supervised classification. A virtual image can be associated with attributes, like real images, but no (real) pixels are associated with it. It is defined as a fixed subsection of a real image, and the corresponding pixels are not stored separately other than in the real image. When a virtual image is stored on (or retrieved from) a disc, only its attributes are stored. It is the responsibility of the programmer to keep track of the associated real image. (This definition of virtual image has not been totally satisfactory to all the group members. We hope to improve it with further experience.)

Syntax:

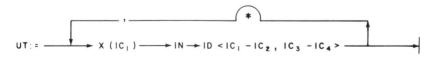

VIRTUAL IMAGE: → [ID_1 IN ID_2 <IC_1 -IC_2, IC_3 -IC_4>] → ; →

(i) ID_1 is the name of the virtual image;
(ii) ID_2 is an already defined real image;
(iii) ($IC_1 - IC_2$), ($IC_3 - IC_4$) defines a rectangular window in ID_2.

e.g. VIRTUAL IMAGE: [X IN A<10:100,10:200>],
                    [Y IN A<100:100,100:200>];

(g) The notion of virtual image group is similar to the image group. It is introduced to experiment with several virtual images and each member of a group is accessed with a subscripted variable.

Syntax:

UT: = → X (IC_1) → IN → ID <IC_1 - IC_2, IC_3 - IC_4> →

where $X(IC_1)$ is the subscripted name of the virtual images and it must be X(1),X(2),X(3)... in the consecutive order. An image group must have at least two members.

VIRTUAL IMAGE GROUP → UT → ;

Since a window does not satisfy all the requirements of masking, the type BOOLEAN IMAGE is defined for the purpose of masking. Each pixel of a Boolean image can be either 1 or 0. It is defined just like images except the keyword BOOLEAN is prefixed. It is possible to define both *Boolean images* and *Boolean image groups*. Both of them are assumed to have *only one band*.

e.g. BOOLEAN IMAGE: P,Q<100,200>;
　　　BOOLEAN IMAGE GROUP: L(3), M(2)<40,50>;

The Boolean constants 1 and 0 are referred to as in ALGOL. Now we have the following image types:

	Real	*Boolean*	*Virtual*
IMAGE	Yes	Yes	Yes
IMAGE GROUP	Yes	Yes	Yes

e.g. VIRTUAL IMAGE GROUP: [X(1) IN A<10−20,10−30>,
　　　　　　　　　　　　　　　X(2) IN B<5−15,5−25>],
　　　　　　　　　　　　　　　[Y(1) IN A<20−40,20−40>,
　　　　　　　　　　　　　　　Y(2) IN C<15−25,15−30>,
　　　　　　　　　　　　　　　Y(3) IN B<0−5,0−8>];

In the above example two virtual image groups X and Y are defined with two and three components respectively. At the execution time, a virtual image (group) is defined only if the corresponding real image (images) is defined and this is the responsibility of the programmer.

In the declaration section, the declarations should occur in the following order:

IMAGE, IMAGE GROUP, VIRTUAL IMAGE,
VIRTUAL IMAGE GROUP and ASSOCIATE ATTRIBUTE.
WINDOW and BOOLEAN IMAGE

Declarations can be placed anywhere in that section.

3. ADDRESSING

In this section we describe how the various data types and their components can be addressed while programming. The three groups of parentheses—(), [], <>—are used as follows:

　(a)　A(2)—refers to the second component image of the image group A.

(b) B[3]—refers to the third band of the multi-band image B.

(c) C<4,5>—refers to the pixel on the 4th row and the 5th column on the band (or single-band image) C.

e.g. some valid combinations are:

A(2)[3]<4,5>—refers to a pixel;

A(2)[3]—refers to a band;

A(2)—refers to an image;

A(2)<4,5>—refers to a vector consisting of the pixel <4,5> in every band of the image A(2).

(d) X .ALL—refers to the image X and all its associated attributes. Only I/O operations can be made with X .ALL.

(e) X .ATT—refers to the set of all attributes of the image X. Like (d), it can be used only for I/O operations.

(f) X:W—refers to the image X viewed through the window W. Before using this mode of reference, the programmer must POSITION the window W on X (refer to section 4).

(g) X .Attr—refers to the attribute Attr associated with the image X .Attr$\in\Omega$, e.g. X .HISTO.

(h) CHANGED (X .HISTO)—is a predicate, TRUE or FALSE (HISTO can be replaced in this example by any Attr$\in\Omega$). Each attribute has associated to it a *status bit*. The status bit is tested by this predicate. CHANGED IS TRUE if the image has been altered, since the last computation of the attribute, in such a way that the attribute no longer corresponds to the modified image.

(i) COMPUTE (X .HISTO)—invokes the computation of the attribute HISTO on the image X and stores the result. Also it resets the status bit. The status bit is set when the image is *changed* during the execution.

(Note that we define a null pixel as a pixel with no value and represent it as Λ. It is different from a pixel with 0 value.)

4. OPERATIONS ON DATA TYPES

The following factors are important:

(i) No operation on image groups is provided; that is, there is no operation similar to the vector addition of APL.

(ii) The addressing modes X .ALL and X .ATT can be used only with I/O operations.

(iii) A virtual image reference can never occur on the left-hand side of an assignment statement. This means pixels (images) cannot be changed by referring through a virtual image.

4.1 Assignment operations

The syntax is similar to that of Algol. Consider the following examples:
A,B,C are images; W1,W2 are windows:

(a) A := B   => $a_{ij} = b_{ij}$ for all i,j.

         A and B must be of same size.

         or $A \geqslant B$ in both dimensions (x,y) and in bands
(see Fig. 1a).

(b) B := C:W1   => The image C viewed through W1 is assigned to B
(see Fig. 1b).

(c) A:W1 := B:W2 => The receiving image A is modified by a window
(see Fig. 1c).

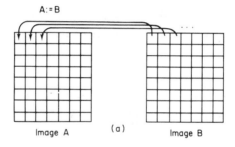

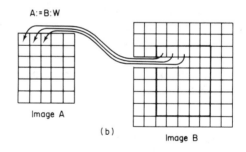

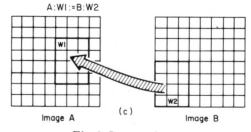

Fig. 1. Image assignment.

4.2 Window operations

A window is treated as a pair of integers defining its frame size. The frame size might vary during the execution of a program.

(a) W := <20,10>
    W := <p,v> —define the frame size for W.

(b) W := W + <1,2>—increments the frame-size by 1 pixel along x, and by 2 pixels along y direction.

(c) XSIZE (W)—is an integer function that returns the frame size of W along the X axis.

(d) YSIZE (W)—is similar to (c) but along the Y axis.

(e) POSITION (W,A,x,y)—will position the window W on the image A at (x,y).

(f) INRANGE (W,A)—is a predicate, TRUE when the window W is completely in the range of the image A (see Fig. 2); otherwise it returns a value FALSE.

(g) LOCATE (W,A,x,y)—is a procedure that returns the (x,y) present position of the window W on the image A.

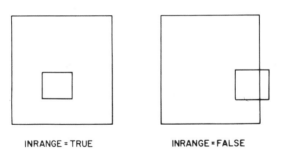

INRANGE = TRUE   INRANGE = FALSE

Fig. 2. Window positions.

4.3 Operations on images

There are two ways to provide operations on images—(i) to define operators as the arithmetic operators +, − etc; (ii) to provide built-in functions as in PAX. We adopt both strategies. For some common and well known operations, we use strategy (i), and for the others we provide built-in functions. Both these aspects are summarized in Table 1 and Table 2 respectively.

Table 1

Operators on images

No.	Operator symbol	An example	Semantics	Remarks
1	$+$	$A = B + C$	$a_{ij} = b_{ij} + c_{ij}$ for all i,j	A,B,C must refer to bands.
2	$-$	$A = B - C$	$a_{ij} = b_{ij} - c_{ij}$ for all i,j	The result is limited between 0 and $2^n - 1$.
3	$\star$	$A = B \star C$	$a_{ij} = b_{ij} \star c_{ij}$ for all i,j	
4	$/$	$A = B / C$	$a_{ij} = b_{ij} / c_{ij}$ for all i,j	Division is integer division.
5	$\cap$	$A = B \cap C$	$a_{ij} = b_{ij}$ iff $b_{ij} = c_{ij}$ otherwise Λ	
6	$\&$	$A = B \& C$	$a_{ij} = b_{ij}$ iff $c_{ij} = 1$ otherwise Λ	C is a Boolean image.
7	$\neg$	$A = \neg B$	$a_{ij} = \max - b_{ij}$	$\max = 2^n - 1$.

5. PREDICATES FOR FLOW CONTROL

Predicates in programming languages are useful to control the flow of execution. They are used in decision making (IF...THEN...ELSE...) and in iterative execution (FOR, REPEAT and DO WHILE statements). In the earlier sections of this report we defined some predicates and they are re-stated here for the sake of completeness.

(a) CHANGED (X.HISTO)—to test the status bit of an attribute (see 3(h)).

(b) INRANGE (W,A)—to test the position of a window (see 4.2(f)).

(c) $<i,j>$ IN A—is TRUE if the pixel $<i,j>$ is inside the image A; FALSE otherwise. The image A may be qualified by a window as A:W. A variant of this predicate is quite useful in a FOR statement as shown below:

FOR ALL $<i,j>$ IN A:W DO;

The following block is executed for every pixel in the image A viewed through the window W.

The order of consideration of pixels is not known to the user. If the programmer needs a specific scan sequence (row major or column major) he should not choose this construct; instead he can construct a (two-level) nested loop. Because the scan-order is insignificant in this construct, asynchronous parallel processors may be used to perform the functions specified [7].

Table 2

Built-in functions

No.	Function name	List of arguments	An example usage	Semantics and remarks
1	MAX	A band	Y=MAX(A[1])	Returns the maximum value of pixels in the band A[1].
2	MIN	A band	Y=MIN(A[1])	Similar, but returns the minimum.
3	MEAN	A band	Y=MEAN(A[1])	
4	VAR	A band	Y=VAR(A[1])	Returns the variance.
5	SHIFT	A band, an integer, a code	SHIFT(A,X,5)	Shift the band A along X by 5 positions. X is replaced by R for right, L for left, U for up, D for down. Vacated pixels are replaced by Λ.
6	ROTATE	A band, an integer, a code	ROTATE(A,X,3)	Similar to shift but no pixel is lost.
7	NUMZ	A band	Y=NUMZ(A[1])	Returns the number of zero pixels in A[1].
8	NUML	A band	Y=NUML(A[1])	Returns the number of null pixels.
9	BOOL	A band, a threshold	A=BOOL(B,τ)	A must be a Boolean image $a_{ij} = 1$ iff $b_{ij} \geq \tau$ $= 0$ otherwise
10	CONECT	A band, a logical variable, two integer variables, an array, a code	CONECT(A,X, p,q,B,L)	In A, a closed area is searched; if found L is set to TRUE; otherwise L is FALSE. The array B contains the Freeman chain of the boundary and (p,q) contains the starting address of the chain. Set X=1 for 4 connectedness. Set X=2 for 8 connectedness.
11	CONERA	Same as CONECT	CONERA(A,X, p,q,B,L)	Finds, the closed region and sets all the interior and boundary points to Λ.
12	CONBON	Same as CONECT	CONBON(A,X, p,q,B,L)	Same as CONERA but the boundary pixels are not replaced by Λ.

Syntax:

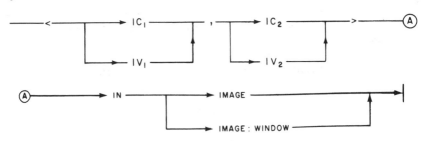

e.g. FOR ALL <i,j,> IN A:W DO;
    IF <1024,2047> IN B THEN PRINT, LARGE;

A number of other predicates can be formed using the functions in Table 2 and the relational operators ($> \geq < \leq = \neq$) of ALGOL.

6. INPUT/OUPUT OPERATIONS

We conform to the syntax of ALGOL, except the I/O list may include images. The I/O list may include any of the following:

$$X, X \text{'ALL}, X \text{'ATT}, X \text{'Attr}, X:W, Y \text{'ATT}, A \text{'Attr}$$

where X is a real image, Y is a virtual image, W is a window and $Attr \in \Omega$. Recollect that when a virtual image is written (or read) only its attributes are written.

As in any I/O operation, it will be the programmer's responsibility to keep track of the record length and the number of items written, etc. For the sake of I/O, the bands of an image will be written in the ordinal sequence 1,2,3 . . . .etc.; each band is written (read) just like a two dimensional array.

As a convention, the attributes of an image will be written (read) before its pixels. The order in which the various attributes are written will correspond to their order in the declaration.

7. ATTRIBUTES OF IMAGES AND BANDS

As stated in section 2, the set of all attributes known to the compiler of L is denoted by Ω. There is only an operational difference between functions and attributes. Certain functions, in PR applications, are intimately associated and used with images. It is worthwhile that such functions may be stored, retrieved, and addressed as an integral part of the image. As a result we have introduced the concept "attributes". An attribute can be associated to one or more images but this association is static. Similarly an image may be associated with one or more attributes.

An attribute is characterized by the following:
(i) Each attribute has a corresponding procedure (function) that computes certain values. The function has only one image as its argument. The COMPUTE (A .HISTO) statement (refer 3(i)) invokes a call to this function.
(ii) The data type returned by the above function determines the storage requirement and a knowledge of this requirement is necessary at compile-time for storage allocation.
(iii) Each association of an attribute with an image has a *status bit* assigned. In sections 3(h) and 3(i), we described the access and control of this status bit. When there is a change in the image, some of the status bits (not necessarily all of them) associated with that image will be *set*. A status bit is *reset* by the associated procedure. Thus, the logic of resetting is embedded into the procedure and the logic of setting is integrated with the semantics of the assignment statement.

The concept of attributes is expected to improve the readability of PR-programs and the in-built documentation. This premise has to be verified in the "language-evaluation stage". However, the programmer should remember that this facility is not without its price.

In Table 3 we present a set of attributes (set Ω) as construed at the time of the

Table 3
Attribute set

No.	Attribute name	Data type returned	Semantics	Remarks
1	HISTO	A vector for each band	Computes the histogram of pixel values of a band	The vector has 2^n elements where n is the number of bits/pixel.
2	MEAN	A scalar for each band	Self explanatory	—
3	Variance	A scalar for each band	Self explanatory	—
4	Covariance	A matrix	Self explanatory	—
		The following are non-computed. Id Info.		
5	Name	Alphanumeric		Name of the image.
6	Date	Numeric	DD/MM/YY	
7	Creator	Alphanumeric		
8	Father-image	Alphanumeric		Of which this is a sub-image.

design of L. After the evaluation stage, this set may change. Eventually, when the system programmer wants to add a new member to Ω, the above requirements must be remembered.

8. PROGRAMMING EXAMPLES

A few examples follow which give an adequate feeling of the expressive power of the language. The simple programs presented implement algorithms which would be much more complex in plain ALGOL.

8.1 Mean image

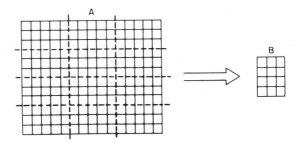

Fig. 3. Mean image.

In this example image B is formed by taking each pixel as a 5 × 3 mean of pixels in image A.

```
IMAGE A<15,12>,B<3,4>
WINDOW V;
V := <5,3>
FORALL <I,J> IN B DO
   BEGIN
      POSITION(V,A,1*5,J*3);
      B<I,J> := MEAN(A:V);
   END;
```

8.2 Image transformation

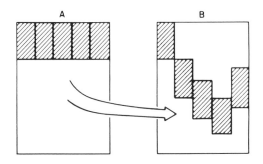

Fig. 4. Image transformation.

This example shows an arbitrary piece-wise transformation from image A to image B which could be useful in geometric correction problems.

```
IMAGE A,B<n,m>;
WINDOW V,W;
V := W := <p,q>;
FORALL <i,j,> IN A DO
  BEGIN
    POSITION(V,A,I,J);
    POSITION(W,B,F1(I),F2(J));
    B:W := A:V;
  END;
```

F1 and F2 are the X and Y transformation functions.

8.3 Correlation

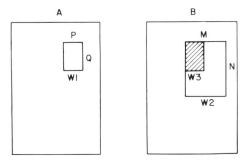

Fig. 5. Correlation.

It is desired to find the position of highest correlation of window W1 within window W2. Therefore W3 (same size as W2) will move within W2 calculating a certain distance function.

```
IMAGE A,B<S,T>;
WINDOW W1,W2,W3;
W1 := <P,Q>;
POSITION(W1,A,X1,Y1);
W2 := <M-P,N-Q>;
POSITION(W2,B,X2,Y2);
W3 := <P,Q>;
M1 := MEAN(A:W1);
FORALL <I,J> in B: W2 DO
   BEGIN
      POSITION(W3,B,I,J);
      M2 := MEAN(B:W3);
      FORALL <K,L> IN B:W3 DO
      DIST := DIST + A:W1<K,L> - M1 % DISTANCE
              - B:W3<K,L> + M2;        % MEASURE
      IF DIST <MINDIST THEN
         BEGIN
            MINDIST := DIST;
            IMIN := I;
            JMIN := J;
         END;
   END:
```

REFERENCES

[1] Guzmán, A. (1979). "Heterarchical architectures for parallel processing of digital images". Tech. Rep. AHR-79-3, IIMAS, Nat. Univ. of Mexico.

[2] Crespi, S. R. and Morpurgo, R. (1970). "A language for treating graphs". *Comm. of ACM* **13**, 319–323.

[4] Johnston. E. G. (1970). "PAX II picture processing system". In *Picture processing and psychopictorics*, Lipkin, B. S. and Rosenfeld, A. (eds). Academic Press, New York and London.

[4] Burroughs B6700/B7700 ALGOL Language Reference Manual.

[5] Burroughs B6700/B7700 Command and Edit Language (CANDE) Manual.

[6] Zahn, C. T. Jr. (1977). "Data structures for pattern recognition algorithms". In *Data structures, computer graphics and pattern recognition*. Academic Press, New York and London.

[7] Guzmán, A. and Segovia, R. (1976). "A parallel configurable LISP machine". Tech. Rep. 133, I, IIMAS, Nat. Univ. of Mexico.

Chapter Four

MAC: A Programming Language for Asynchronous Image Processing

R. J. Douglass

1. INTRODUCTION

The programming language MAC is designed for writing asychronous parallel image processing algorithms. It provides tools for specifying, replicating and activating processes and operations for controlling communication between them. MAC combines the concepts of process, guarded commands, and guarded regions found in existing languages for distributed processing with a new set of language features designed for asynchronous image processing. Although the syntax of the language is similar to the "distributed processes" language (DP) of Brinch Hansen [1], the new language constructs in MAC make it suitable for a very different class of algorithms than previous distributed processing languages such as DP or CSP [2].

The last five years have seen the development of numerous language designs for parallel and distributed processing in general, and parallel image processing in particular [1–13]. Most of the languages for distributed processing [1, 2, 8, 9, 10, 13] are intended for systems of relatively few distinct processes with relatively infrequent communication and asynchronous execution. These languages have developed techniques for handling the control and communication between asychronous processes. In contrast, most of the parallel image processing languages [4, 5, 6, 7] are designed for SIMD array architectures where there are many identical processes executing synchronously with frequent interprocess communication. These languages have shown how parallelism can be expressed and managed as a set of identical operations on each element of a data structure, usually a one or two-dimensional array. It was argued by Douglass [14] that there exists a large class of image processing algorithms which share elements of both distributed processing algorithms and SIMD algorithms. MAC provides a bridge between these two types of parallelism.

2. IMAGE GRAPH ALGORITHMS AND THEIR IMPLICATIONS FOR PROGRAMMING LANGUAGE DESIGN

2.1 Image graph algorithms

Most parallel image processing algorithms are for low-level processing such as smoothing, filtering, and gradient operations, and these algorithms fit very nicely into an SIMD array format. There is, however, a class of image processing algorithms that does not fit the SIMD mould because even though they are highly parallel, they are asynchronous. These algorithms are also distinguished from traditional MIMD algorithms because they use identical processes replicated over each element of a large data structure.

In contrast to low-level image processing algorithms which can be expressed as parallel operations on every point of a two-dimensional array representing an image, higher-level image processing uses a more abstract description of the image, usually in terms of edges or regions. These descriptions can be thought of as a graph where the nodes of the graph represent regions or edges and the links between the nodes represent the connections between neighbouring regions or edges in the image. One class of parallel algorithms processes such image graphs by assigning an identical process to each node of the graph. Each process updates its own node's description using information from the description of neighbouring nodes. Relaxation labelling and region matching for change detection are two examples of this class of algorithms.

2.2 The requirements of a language for image graph algorithms

The requirements of a programming language which can support image graph algorithms were given by Douglass [14] and can be summarized as follows: (i) ability to define identical or similar processes to work on different parts of a large common data structure; (ii) support parallelism at the level of procedures; (iii) permit the dynamic creation and destruction of processes and their interconnections; (iv) allow closely coupled processing; (v) allow simultaneous reading of shared data by several processes, and (vi) enforce sequential modification of shared data.

3. MAC LANGUAGE CONCEPTS

MAC is a set of language constructs for defining cooperating parallel processes and controlling their interactions. MAC meets the six criteria outlined in the preceding section. A MAC program consists of a process definition section followed by declarations of a number of instances and interconnections of those processes. A process can be replicated over each element of a data structure such as an array, and each process can activate other processes. Processes communicate either by calling procedures or functions in another process or by reading shared variables. Processes are assumed to run in parallel on virtual processors and there is no mapping implied between virtual and physical processors.

3.1 Processes

The central element of a MAC program is the process which consists of a process name, a list of variables, a list of subprocesses, a set of procedures and functions, and an initialization section. For example, a semaphore process would be defined as follows:

```
PROCESS pv;
    s : INTEGER;
    PROC signal;
        s := s+1;
    PROC wait;
        WHEN s >= 1 : s := s-1;
    INIT
        s := 0;
    END; {of process pv}
```

"WHEN" is a guarded command which waits for condition $s >= 1$ to become true.

Except for pointer processes described below, processes are created and perform their initialization section at the start of program execution. After they have finished their initialization section or whenever they are forced to wait on a guarded command, processes are free to execute any of their procedures or functions that might be invoked by another process. For example, a user process could perform a wait using the semaphore process by calling pv. wait.

The "PROCs" and "FUNCs" consist of regular sequential PASCAL code plus the communication and control constructs described in Section 3.4 below. A procedure may call another procedure in any other process providing it has access to the process's name.

The process variables exist for the life of the process and are accessible to all procedures in the process. Other processes may read but not directly modify a process's variables using a special READSHARED operation. Variables declared in a PROC are local to that procedure and are not visible outside the scope of that procedure. Subprocesses are declared as instances of previously defined prototype processes as described in the next section.

3.2 MAC programs

MAC programs are a collection of asynchronously executing processes. A program consists of a PROCESSTYPE section which defines prototype processes followed by a PROCESSVAR section which declares actual instances of those processes. When a program begins execution, all processes declared in the PROCESSVAR section are activated and run in parallel. The PROCESSTYPE section is the analogue of the TYPE section of a PASCAL program. In the PROCESSVAR section a data structure may be defined which replicates a prototype process over each element of the structure. For example, an array processor similar to the Massively Parrallel Processor (MPP) [15] can be defined as a two-dimensional array of simple process elements:

```
PROCESSTYPE
    PROCESS element;
        memory : ARRAY [0..1023] OF 0..1; {PE element memory}
        G,A,B,P : 0..1; {PE registers}
        PROC
            {procedures and functios to define the basic operations on each
                                                              element}
        INIT
            {initialization section for a PE: zero memory, set G := 1;}
        END; {of process element}
PROCESSVAR
    PROCESS mpp : ARRAY [0..127,0..127] OF PROCESS element;
END.
```

Processes can be associated with elements of a data structure by directly dimensioning a vector or array of identical processes when they are defined in the PROCESSVAR section. For example, a pipeline of ten identical processing elements could be declared as follows:

```
PROCESSVAR
    PROCESS pipe[1..10]; {definition of an individual pipeline element}
```

A process may refer to its neighbours using the intrinsic function SELF plus an integer offset. Thus, a particular pipe process can address its left neighbour as pipe[SELF − 1]. The intrinsic function SELF is used in the Finite Element Machine as well as the DP language [1].

3.3 Pointers to processes

The preceding section showed how a process can be duplicated on each element of an array, but parallel image graph algorithms require a language which permits the definition of more general interconnection patterns than simple arrays. To meet this need, MAC provides the ability to define pointers to processes. Just as process arrays in MAC are analogous to arrays of data in PASCAL, MAC process pointers are analogous to PASCAL pointer data types.

As an example of pointers to processes, consider a process "createregions" which scans an image segmenting it into regions of approximately uniform colour and texture:

PROCESS createregions;
   numofregions : 0 . .maxregions;
   regions : ARRAY[1 . .maxregions] of ↑ PROCESS region;
   {the definition of PROCs and FUNCs and the INIT section of create-
                                          regions goes here}
END; {of process createregions}

Whenever "createregions" forms a new region description, it activates an instance of process region by a call to the intrinsic operation ACTIVATE: ACTIVATE(regions[numofregions+1]). In this fashion, each region gets its own process which will be responsible for updating the description of that region. Unlike the usual segmentation algorithm which returns a data structure containing region descriptions, the result of createregions is the formation of a network of running processes. Section 3.4 shows how such a network can be used to match stereo pairs in parallel.

Processes defined as process pointers such as the individual region processes described above are not initially active. Only when an ACTIVATE operation is performed is a pointer process created and its initialization section executed. References to pointer process are made using the ↑ notation as in PASCAL for data pointers. For example, a procedure in the ith region process could be invoked by regions[i] ↑ .procname, assuming the process regions[i] ↑ had already been activated. A given process may be pointed at by another process pointer variable. The sequence:

ACTIVATE(regions[1]);
ACTIVATE(regions[2]);
regions[2] := regions[1];

will activate two processes but will leave the process pointer variables regions[2] and regions[1] referencing the same process. The process created by ACTIVATE(regions[2]) will still exist, but regions[2] will no longer refer to it.

3.4 MAC control constructs

MAC uses the control constructs listed in Table 1. The key control statements are FORK, ACTIVATE, READSHARED, and the generalizations made on Brinch Hansen's guarded commands and regions. The last section discussed the definition of prototype processes and pointers to processes and the ACTIVATE statement which is necessary for creating a new instance of a pointer process. FORK, READSHARED, and guarded commands and regions are constructs which permit communication and synchronization between processes. The "FOR x IN a : s" statement is identical to the FOR statement in DP which sequences x in statement s through each element of a structured data item a. The IF, DO, WHEN and CYCLE statements are extensions of the non-deterministic guard statements introduced by Brinch Hansen.

Table 1
MAC control constructs

Guarded commands: IF (ELSE) DO IF ANY (ELSE)	Reading shared data: READSHARED
Guarded regions: WHEN WHEN ANY CYCLE	Process invocation: FORK CALL function calls
Process creation/termination: ACTIVATE TERMINATE	Sequencing: FOR

As in DP, the IF statement enables a process to select between several statments by inspecting the state of its variables. If none of the alternatives are true, the IF statement is skipped or a program exception occurs. In MAC an ELSE clause may be specified along with the IF. The IF statement has the following syntax: IF $b_1 : s_1 | b_2 : s_2 | \ldots$ ELSE s END. The b_i represent boolean expressions, and the s_i are statements. If one of the b_i is true then the corresponding s_i is executed. The IF statement is non-deterministic because an arbitrary choice is made if more than one b_i is true. The ELSE part is optional.

Since MAC provides functions, it is possible to have an IF statement where the b_i are calls to boolean functions in other processes. Whenever a condition b_i

is a function call, MAC will initiate execution of the function and then move on to evaluate the next condition without waiting for the completion of the function. Thus an IF statement can be used to initiate the parallel execution of several functions. If any function returns a true value, the corresponding statement will be executed. If all functions return a false value, then the statement will be skipped. A DO statement is the endless repetition of an IF statement.

The WHEN statement is identical to the IF except that a WHEN statement causes the process to wait at the WHEN if none of the conditions are ture, and to remain blocked repeating the test conditions until one of them becomes true. As with the IF statement, a WHEN statement will initiate concurrent execution of any Boolean functions that are part of its test conditions. A CYCLE statement is an endless repetition of a WHEN statement.

MAC provides the WHEN ANY and IF ANY statements as a shorthand way of performing a WHEN or IF on each element of a process array. For example, given an array of processes: a : ARRAY[1 . .n] OF PROCESS p and a Boolean function b in process p, an IF statement executing all *n* functions in parallel would be:

IF ANY x IN a: x .b:—execute a procedure or function in x.

The FORK statement in MAC is used to initiate the concurrent execution of a procedure in another process. For example, a process p can call a procedure r in another process q in one of two ways: (i) CALL q .r(parameter list); or (ii) FORK q .r(parameter list);. In the first case, process p will call procedure r and wait until it has finished execution and returned control to p. In the case of the FORK, execution of procedure r is initiated by the FORK statement but process p continues executing concurrently with q .r.

The final control construct introduced by MAC is the READSHARED statement which has the following syntax:

READSHARED processname : statement;.

The purpose of a readshared statement is to permit a process to directly examine the values of the variables in another process. READSHARED enables a level of parallelism in some algorithms such as stereo matching that would not be possible without shared reading of common data.

Since variables can be modified only by the process that defined those variables, modifying variables is kept strictly sequential. A process must be sure that no other process is reading its data when it modifies that data. Prefixing the reading of shared data with a READSHARED operation ensures that all attempts to read shared data are gated through a semaphore variable associated with that data. The semaphore variable allows a process to lock out the reading of its variables when it is modifying them.

4. TWO EXAMPLES OF PARALLEL ALGORITHMS IN MAC

4.1 Associative retrieval

As a first example of a MAC program, consider an associative memory which is defined as a vector of processes where each process is responsible for searching and updating a portion of the associative memory. Such a process can be defined in MAC as follows:

```
PROCESSTYPE
  PROCESS cell; {a single retrieval element of the assoc. memory}
      memory : ARRAY[1 . .10] OF dataitems;
          {each cell will search or update its own ten memory elements}
      founditem : dataitem; {used to return an item found during a memory
                                                                    search}
      FUNC search (key:keytype) : boolean;
          {a function to search for an item matching a given key}
      VAR x:dataitem;
      BEGIN
          FOR x IN memory :
              IF key=x .key :  BEGIN
                                  founditem : = x;
                                  search := TRUE;
                               END
              ELSE search  := FALSE;
      END; {search}
PROCESSVAR
  PROCESS associative;
      assocemem:ARRAY[1 . .100] of PROCESS cell;
      x .PROCESS cell:
      PROC recall(key:keytype#item:dataitem);
          {recall returns the item in "item" corresponding to key}
      BEGIN
          IF ANY x IN assocmem:
              BEGIN
                  x .search(key);
                  READSHARED x : item := x .founditem;
              END
          ELSE item := nil;
      END;
      INIT
          {initialization: read in the initial data for each cell of the memory}
      END; {of process associative}
```

4.2 Parallel stereo region matching

Deriving depth information from stereo images requires an algorithm that can match up corresponding parts in the two images. The algorithm given below assumes that both images have been segmented into region descriptions and that the "createregions" routine given in Section 3.3 has associated a process with each region in one of the images (image A). Each region process in image A first creates a list of regions from image B as candidates for a match. Next, the process region initiates a matching process for each candidate which will return a value expressing the goodness of fit between the region in image A and the given candidate in image B. Finally, procedure "selecthighest" will select the candidate with the highest matching score as the corresponding match for the region in image A. Process createregions has activated a graph of processes of the form: regions : ARRAY [1 . .numofregions] of ↑ PROCESS region, where PROCESS region is defined in the PROCESSTYPE section as:

```
PROCESS region;
    matchingregion : INTEGER; {index of the corresponding region in
                                                    image B}
        candidate : ARRAY[1 . .maxcan] OF candidatedescriptions;
            {an array holding a list of regions which are candidates for
                                                    matching}
        numcan : 0 . .maxcan; {the actual number of candidates found}
        matcher : ARRAY[1 . .maxcan] OF ↑ PROCESS match;
                        {process match is defined below}
    PROC createcandidates;
        {this procedure selects 0 to maxcan regions from image B as
                            candidates for a match with this region
                            and stores them in candidate}
    PROC selecthighest;
        {sets matchingregion to the candidate with the highest match value}
    PROC domatch;
        VAR i : INTEGER;
        BEGIN  FOR i IN [1 . .numcan];
                    BEGIN ACTIVATE(matcher[i]);
                    FORK(matcher[i] ↑ .compare(self,candidate[i]#
                                            candidate[i] .metric);
                        {this FORK initiates the concurrent matching of
                                    this region to each of candidates
                                    in image B.}
                END;
        END;
```

```
INIT
    {initialization of process region}
END; {process region}
PROCESS match;    {match makes the comparison between two regions
                             using the PROC compare}
    PROC compare (reg1,reg2:regionindices#matchscore:INTEGER);
                  {compare scores the match between regions reg1 and
                             reg2 and returns a score in matchscore}
    INIT
END; {of process match}
```

5. CONCLUSION

MAC combines the SIMD view of parallelism where identical processes operate synchronously on each element of a large data structure with the distributed processing view which provides constructs for defining and controlling asynchronous processes. The result is a new language for expressing image graph algorithms: algorithms which are neither wholly SIMD nor MIMD in form but combine elements of both. MAC is intended as a language for investigating parallel image processing algorithms rather than a complete language specification ready for writing production programs. MAC and languages like MAC will increase our understanding of parallel algorithms by extending our ability to express those algorithms.

REFERENCES

[1] Brinch Hansen, P. (1978). "Distributed processes: a concurrent programming concept". *Comm. ACM* **21,** 934–940.

[2] Hoare, C. A. R. (1978). "Communicating sequential processes". *Comm. ACM* **21,** 666–677.

[3] Perrott, R. and Stevenson, D. (1978). "ACTUS—a language for SIMD architectures". Proc. of LASL Workshop on Vector and Parallel Processors, Los Alamos.

[4] Uhr, L. (1979). "A language for parallel processing of arrays embedded in PASCAL". Comp. Sci. Tech. Rep. 365, Univ. of Wisconsin.

[5] Reeves, A. (1980). "Parallel PASCAL". Int. Conf. Parallel Processing, Boyne Highlands.

[6] Barrera, R., Guzmán, A., Ginich, A. and Radhakrishan, T. (1979). "Design of a high level language for image processing". Tech. Rep. PR-78-22, IIMAS, Nat. Univ. of Mexico.

[7] Levialdi, S., Maggiolo-Schettini, A., Napoli, M., Tortora, G. and Uccella, G. (1981). "Preliminary results on the construction of an image processing language: PIXAL". (This volume.)

8] Brinch Hansen, P. (1975). "The programming language Concurrent PASCAL". *Trans. IEEE Soft. Eng.*, 1975, 199–206.

9] Wirth, N. (1977). "Modula: A language for modular multiprogramming". *Software— Practice and Experience* 7, 3–35.

0] Cook, R. (1979). "*MOD—A language for distributed programming". First Int. Conf. Distributed Processing, pp. 233–241.

1] Ison, R. (1977). "Parallel processing for a polyprocessor computer". Tech. Rep. TR-13-77, Dept. of Applied Maths and Comp. Sci., Univ. of Virginia.

2] Reynolds, P. (1980). Ph.D. Thesis, Univ. of Texas.

3] Feldman, J. (1979). "High level programming for distributed computing". *Comm. ACM* 22, 353–367.

4] Douglass, R. (1980). "The requirements of a language for asynchronous parallel image processing". Int. Conf. Parallel Processing, Boyne Highlands, pp. 147–148.

5] Batcher, K. (1979). "MPP—A massively parallel processor". Int. Conf. Parallel Processing, pp. 249–250.

Chapter Five

A Language for Parallel Processing of Arrays, Embedded in PASCAL

L. Uhr

1. INTRODUCTION

This chapter describes an experimental language (called PascalPL) that extends PASCAL with constructs to process parallel arrays. PascalPL allows the programmer to co.!e procedures that effect parallel operations over arrays of integer (or, optionally, boolean) values of the sort commonly used in image processing, pattern recognition and scene description programs. Statements can specify sets of relative locations (with respect to each cell of the array), and compounds of such sets, embedded in assignment statements, or in conditional statements.

PascalPL does not ¡.urport to handle efficiently a wide range of numerical operations. Nor does it attempt to handle the problems of parallel-serial processing by networks in general. Rather, arrays of information (e.g. as input by a telelvision camera, and transformed by arrays of hardware (or virtual) processors) are the major type of parallel structures that it addresses. But since it is embedded in PASCAL all the facilities and power of PASCAL are available to the programmer.

PascalPL presently exists as a program (coded in PASCAL and running under the UNIX operating system on the University of Wisconsin Computer Sciences Department's VAX) that inputs a legal PascalPL program and outputs a legal PASCAL program that contains the code that the PascalPL pre-processor outputs (which will now effect the parallel operations called for by the PascalPL code). Its special parallel constructs (to be described below) include parallel assignment (||set .. := ..;), conditional (||if .. ||then .. ||else ..;), input (||read(..);) and output (||write(..);) statement-types. These all execute operations, in true parallel fashion, on arrays of information. This is effected on the VAX, as on any other serial computer with only one processor (the "central processing unit" or "cpu"), by setting up temporary arrays (to keep processes parallel) and embedding the processes within nested do-loops. (See the Appendix for examples.)

2. THE NEED FOR HIGHER-LEVEL LANGUAGES FOR PARALLEL ARRAYS

Several large arrays of parallel processors are today just beginning to become available. CLIP4, a 96 × 96 array [5, 6], has each of its almost 10,000 processors fetch, in parallel, information from any subset of its eight nearest neighbours plus its own memory and execute a logical instruction. ICL's DAP [8, 20] is a 64 × 64 array with a good bit of capability for numerical as well as for image processing problems.

MPP (designed for NASA by Goodyear-Aerospace, to be delivered in 1982; see [9]) will be a very fast 128 × 128 array. Goodyear-Aerospace is also building ASPRO, a 2000 × 1 reconfigurable array with 32 processors and a flip network on a single chip, for delivery by 1981 [1]. (ASPRO is closely related to Goodyear's Staran-E [2], essentially using lsi technology to miniaturize a STARAN into a one cubic foot box.) Two pipelined and specially designed computers have also been built for image processing: PICAP [14, 15], and the CYTOCOMPUTER [23].

When actual array-processing hardware is connected to a serial computer that executes PASCAL programs then the parallel constructs in PascalPL can be translated directly to invoke the parallel hardware, which will execute them directly (and with two to five orders of magnitude increases in speed).

For the present it seems of interest to begin to develop parallel languages, to ease the burden of coding parallel programs (even if for a serial computer), to explore what kinds of parallel constructs are useful, and to examine what mixes of parallel and serial constructs programmers actually use.

It is not at all clear whether present-day languages that have been developed for serial computers should be extended to handle arrays and networks of parallel processors, or whether entirely new languages should be developed. In the long run we will inevitably see entirely new languages, that reflect the new hardware architectures of computers as well as our advances in understanding language design. But for the next three or five (or ten?) years it seems likely that the most powerful systems will use a serial computer (to serve as "host", file manipulator, compiler, simulator, etc., as well as to execute the serial portions of programs) along with any hardware arrays and networks of processors. It therefore seems appropriate to take advantage of the serial computer, and of the power of the newer languages that have been developed for serial computers.

PASCAL [26, 13] seems the best choice for the language-to-be-extended. It is already widely used and promises to be used increasingly, especially on microprocessors of the sort used to build networks and arrays. It is both powerful and relatively efficient. It has been extended, in concurrent PASCAL [3] and MODULA [27], to begin to handle networks of processors. It is being extended, in TELOS [24] to handle a wide range of artificial intelligence and data base management tasks. At the first of what is expected to be a series of annual conferences on parallel architectures and higher-level languages for image processing, the consensus of those who felt a modern language was needed was that it should be embedded in PASCAL (until a completely new language emerges), and a working group was formed with that aim in mind [7]. Finally, since it seemed desirable to implement an experimental version of such a language, to explore its problems and its uses, a decision had to be made as to the specific language to extend. PASCAL seemed a reasonable choice.

3. A BRIEF BACKGROUND REVIEW OF RELATED WORK

A large number of languages have been coded to handle parallel array processes [19]. These have tended to be at the assembly language level, or to be sub-routine calls in FORTRAN. But they include a number of languages that are quite simple and straightforward to use, and offer the user great power (to a large extent because the parallel arrays, with their very large numbers of processors, are so powerful). Most of these languages execute entirely on a particular machine with parallel hardware (or on a simulation of parallel hardware), and are directed toward image processing. (The languages that have been developed for array processing of numerical problems, e.g. APL [12], GLYPNIR [16], and ACTUS [18] for the ILLIAC-IV super-computer, will only be mentioned here.)

Kruse and his associates [10] have developed a very nice ALGOL-like language in which programs can be coded that call both the PICAP parallel processor and the serial host computer. It contains two major parallel constructs, for (i) logical, and (ii) numerical operations, with the format:

$$\text{(type) } \begin{array}{ccc} 1 & 2 & 3 \\ 4 & 5 & 6 = \text{result} \\ 7 & 8 & 9 \end{array}$$

That is, the programmer is asked to specify a 3×3 array of nine values (these may be boolean values, or integers specifying weights or inequalities), plus an operation and a result. This closely reflects the nearest-neighbour structure of the PICAP-1 array (and of most other of today's hardware arrays).

Uhr [25] has developed a higher-level language (also called PascalPL, but let's refer to it as PascalPL .0) for the CLIP parallel array that makes use of "compounds" and "implications" similar to, and precursors of, some of the constructs in the present PASCAL-based PascalPL.

Reeves [21] has developed a system that extends APL to handle image processing as well as numerical processing of arrays. He argues [personal communication] that APL, since it handles arrays quite naturally, may well be preferable to PASCAL as the "host language" into which parallel constructs should be embedded. But since PASCAL is far more widely available, and far more consonant with the feelings that most people have today about what is "good structure", it seems the better choice. APL is, however, a language that can suggest some useful constructs that might be embedded in a PASCAL-based system.

Levialdi and his associates [17] are developing PIXAL, which consists of parallel extensions embedded in ALGOL-60 (which they are using chiefly because PASCAL is not available on their computer). PIXAL uses a "frame" construct that allows the programmer to specify a set of relative locations (not only nearest-neighbours) to be looked at everywhere (that is, relative to each cell in the array), and a "mask" construct with which the programmer can specify a set of weights to be applied to the relative locations specified in the co-ordinate structure.

Douglass [4] has proposed exentions to PASCAL to handle parallel arrays, building on earlier proposed extensions by Pratt and Ison [11] to handle networks. These include a "fork" operation to set up new processors that are executing different sets of instructions in parallel, and a "split" operation, to invoke whole sets (e.g. arrays) of processors to execute the same set of instructions, but each on different data (e.g. different local regions of a visual image). The programmer can specify "windows" (sets of (relative) co-ordinate locations).

Schmitt [22] has also described extensions to PASCAL to define and then execute operations on any arbitrary network structure, rather than limiting the programmer to arrays.

Almost all of these languages use as their key construct an operation on a set of relative locations. The languages designed for a specific hardware array will build in the interconnection pattern of that array (usually, as in the case of PICAP-1, the 3×3 sub-array of the nearest neighbours). PIXAL's "mask" and Douglass' "window" generalize this to any set of arbitrarily distant neighbours. (But when such a language is actually executed on a hardware-parallel array this gives excessively long sequences of nearest-neighbour shifting operations.) Uhr added constructs for the convenient compounding and implying of information over several different arrays. Douglass and Schmitt make suggestions for constructs to handle much more general networks, as well as SIMD (Single-Instruction-Multiple-Data-Stream) arrays.

4. A DESCRIPTION, WITH EXAMPLES, OF PASCALPL

PascalPL is designed to handle arrays, and sets of arrays, that contain binary, grey-scale and/or numerical values. It allows one to code procedures that look at and compound sets of relative locations (around each "pixel" cell) in a single array, and to compound sets of these sets across several arrays, using either arithmetic or boolean operations.

It seems best to introduce the reader to PascalPL by starting with very simple examples, gradually introducing its full set of constructs and features.

4.1 An overview of PascalPL constructs

A PascalPL program looks like a PASCAL program, except that it contains several new constructs (all announced and made visible by two vertical [parallel] bars, e.g.: ‖procedure..; ‖set..; ‖if..; ‖read..; ‖write..;). This means that the programmer must declare all the necessary constants, data types and variables, and strictly follow all the conventions of PASCAL (e.g. a program must begin with "program{programname}(input,output{...});" and end with "end."). The programmer can code as much as she/he desires in ordinary PASCAL. Only when parallel constructs are desired do any deviations occur. These parallel constructs are handled as follows:

The procedure that will contain these (one or more) parallel constructs must be declared:

‖procedure {procedurename};

Any time after this procedure's "begin" statement, a "dimension declaration" statement must be placed, e.g.:

‖dim [0..127,0..127];

Then come, interspersed with ordinary PASCAL statements, parallel constructs of the sort:

‖read(..);
‖write(..);
‖set {array(s) assigned to} := {compound of array(s)..};
‖if{(compound of arrays)(ineq)}
   ‖then {array(s) modified}
      ‖else {arrays modified on failure—optional};

Procedures cannot be nested within parallel procedures (this restriction will be lifted when PascalPL is actually embedded in PASCAL, and can conveniently use stacks for declarations).

4.2 A very simple example program

The following program uses only the simplest of PascalPL constructs. (See the Appendix for the PASCAL program into which the PascalPL preprocessor translates it.)

```
program simple(input,output);
{the programmer must declare the array data structures used}
{for this program they are: image, negative, edgedimage}

procedure sayhello;
begin
   writeln('hello');
end;

||procedure demonstrate;
begin
   ||dim [0..2,0..2];
      ||read(image);
         ||write(image);
      ||read(negative);
   writeln('the image and the negative image have been input');
      ||set edgedimage := image − negative;
      ||write(edgedimage);
      ||set image, negative := 0;
||end;

begin {program}
   sayhello;
   demonstrate;
end
```

4.3 An example of output from the simple program above

```
hello
TYPE IN INTEGERS FOR image[0..2, 0..2]
      1   2   3   were input to row 0
      4   5   6   were input to row 1
      7   8   9   were input to row 2
ARRAY = image contains:
      1   2   3;
```

```
4   5   6;
7   8   9;
END OF ARRAY.
TYPE IN INTEGERS FOR negative[0..2,0..2]
5   5   5   were input to row 0
5   5   5   were input to row 1
5   5   5   were input to row 2
the image and the negative image have been input
ARRAY = edgedimage contains:
-4  -3  -2;
-1   0   1;
 2   3   4;
END OF ARRAY.
```

This program first outputs "hello" and then inputs a 3 × 3 array, naming it "image". (As it is presently implemented to run interactively, it outputs the message to TYPE an array, and outputs the inputs, to verify.)

The next statement "‖write(image)" outputs the array stored in image, in array form. "‖read(negative)" inputs the array that it names "negative". Now the program outputs that the two arrays have been input. (Note that this is a regular PASCAL "writeln(..);" command. It illustrates how regular PASCAL statements can be interspersed.)

Now, for each cell in the array "edgedimage" the program subtracts what negative contains from what image contains. Next it re-initializes the two arrays image and negative, so that each of their cells contains a zero. Finally, it outputs the array named edgedimage.

This is a trivial example, and does not begin to indicate any of the more powerful ways that the parallel-assignment construct (‖set) can be used. But it shows how PascalPL's parallel constructs can be intermixed with standard PASCAL, and it does show how input and output are extended, in a straight-forward but what appears to be satisfactory way, to handle arrays.

4.4 The assignment statement

An assignment statement contains a set of "arrays-to-be-assigned-results" (called "assignees") to the left of the assignment operator (":=") and a "compound-of-structures" (called "compounds") to the right of the ":=".

Now let's examine a sequence of more powerful assignment statements (see the Appendix for the PASCAL code output by PascalPL for selected statements):

‖set vert := image[+(0:1,0:0,0:−1)];

(This statement looks, for each cell in image, at the 3 relative locations specified, sums what it finds, and stores the result in the corresponding cell of vert.)

$\|$set cross := vert[+(0:0,0:2,0:−2)] * hor[+(0:0,2:0,−2:0)];

(sums the 3 relative locations in vert, does the same for hor, then multiplies these sums and stores the result in cross.)
Note that:

$\|$set array1 := array2 (op) array3;

is equivalent to:

$\|$set array1 := array2[0:0] (op) array3[0:03);)

$\|$set vert, hor, image := 2 * image[+(0:1,1:0)] − #average;

(This illustrates how several arrays can be assigned the values computed by the compound, and how the compound can include constants and variable identifiers [each must be preceded by '#'].)

$\|$set gradient := image[+(0:0*12,1:0*−2,0:1*−2,−1:−1*−1, . . .)];

(This will get a weighted difference between the centre cell (weighted + 12) and the 4 square neighbour cells (weighted −2 each) and 4 diagonal neighbours (weighted −1 each) (only the centre and 3 of the 8 neighbours are shown).)

$\|$set featurei, featurej, labelk
   := featurem3(*(4:−3*2>5,−5:7*3>14,0:0*21>112)];

(multiplies what is found in each relative location by the specified weight and accepts the result only if it exceeds the specified threshold; then stores the sum of these results in the corresponding cells of featurei, featurej and labelk.)

   To summarize, an assignment statement consists of an arbitrarily long compound (to the right of the assignment operator ":=") of array-specifications. An array-specification consists of an array name followed (optionally) by a "structure". A structure consists of an (integer or boolean) operator followed by a set of relative locations. Each relative location in an integer array can, optionally, be followed by an operator and a weight and/or an inequality and a constant.

   One or more arrays can be named to the left of the assignment operator. Each will be assigned the result of the set of operations on arrays specified in the compound. (An option that is probably not very useful also allows (for integer arrays) an arithmetic operator followed by an integer, to modify each result by the specified constant before storing it in its corresponding cell.)

4.5 The conditional statement

A conditional statement can also be constructed, with the form:

‖if {compound} {optional inequality} ‖then {modifications};
  or
‖if {compound} {opt. ineq.} ‖then {modif.} ‖else {modif.},

For example (see the Appendix for PascalPL's PASCAL output):

‖if featurei[+(0:1,0:−1)] * featurej[+1:0,−1:0)] > 11
  ‖then labeli+19, labelj*2, labelk−33
  ‖else labell*2, labelm+27;

The conditional statement first computes the compound (for each cell in the arrays). Then, if an inequality is specified, it tests it and (only when it is satisfied or, when using boolean arrays, if the compound is true) makes modifications to the arrays following the "‖then". If there is also an ‖else, what follows it is modified (only when the conditional fails, or is false).

The modifications can specify an operator and an integer (e.g. labeli+19, indicating "add 19 to labeli", or vert*36, to multiply vert by 36).

4.6 Reading and writing arrays

The two simple constructs:

‖read({arrayname}); and
‖write({arrayname}):

input and output the specified array.

4.7 Constructs that declare parallel procedures

The ‖procedure {procedurename}; construct must declare each procedure that contains one or more ‖if . . . or ‖set . . . statements. (Its purpose is to signal the PascalPL pre-processor that a parallel procedure will have to be set up, so it can declare that procedure here, with a "forward". This avoids any problems that might arise if the programmer used the same name in the declarations that PascalPL uses in declaring the temporary lists and data structures that it must use. If these extensions were embedded in the PASCAL compiler itself this declaration could be eliminated or, probably better, combined with the ‖dimension declaration.)

The ‖dim statement declares the array-type and dimensions of the arrays to be used in the parallel constructs that follow in this procedure. Its shortest and simplest form is:

‖dim;

(which declares the previously declared, or default, dimensions and integer arrays).

Its full form is:

||dim {arraytype}merge from {from-setname}[{arraydimensions}]
    to {to-setname} [{shrink-convergence}];

The arraytype can be & (for boolean) or * (for integer). The from-setname and to-setname are optional. If used, they are concatenated in front of the names of arrays in (a) compounds and in (b) assignees or modifiers, respectively. (This is an option that may be useful when sets of arrays are collected into layers or characteristics, and if and when a feature that lets the programmer declare and use them as arrays of records, using the PASCAL "with" construct, is implemented.)

The array dimensions must be given in the form: [minx . .maxx,miny . .maxy], optionally declared boolean or integer—e.g. [0 . .127,0 . .63 : integer]. (Note that his means the array-type can either be declared before or within the array dimensions. Whichever seems more convenient and more natural will be retained in the future.) Thus still another alternative declaration is:

||dim [0 . .7,0 . .15 : boolean] {optional shrink};

The shrink-convergence indicates how information is converged from a from-array to a to-array. For example, if it is 2,3 then the x-co-ordinate of a from-layer cell is divided by 2, and the y-co-ordinate is divided by 3, to compute the co-ordinates of the to-layer's cell. (If no shrink-convergence is specified the program will use whatever was last specified. The default, if shrink-convergence was never specified, is 1,1—that is, co-ordinates are divided by 1, so that no convergence occurs.)

An end statement:

||end;

ends the parallel procedure.

The type of border to be used when a relative location-to-be-looked-at lies outside the array is specified by:

||border := {bordertype};

where bordertype can equal 0 (for "false"); or 1 (for "true"); or 2 (for "what is contained by the nearest cell within the array"). (The default condition is border = 2.)

5. SUGGESTIONS FOR POSSIBLE FUTURE EXTENSIONS

As noted above, PascalPL can be simplified and improved when it is embedded within PASCAL itself (this could have been done in the present pre-processor, but did not seem worth the effort in a first experimental system). The symbol table PASCAL builds up could be used to merge the names PascalPL must declare with those that the programmer declares (and also change names that are the same). PASCAL could get each array's types when it is declared; then PascalPL could generate the appropriate temporary arrays and stores (but it may be best to limit the programmer, since there will rarely be more than one physical hardware array present).

If declarations of dimensions, border-types, array-types and shrink-conversion were stacked parallel blocks could be embedded, and treated exactly like ordinary blocks (but this would violate the capabilities of most parallel hardware). Procedures could be generated and then called, to shorten the code (but the present code seems more appropriate for a first experimental system, since it makes clear exactly what serial PASCAL code must be executed to effect each parallel PascalPL procedure). Code could be made more efficient in several places. For example, the temp{orary}store is often not needed; the conditional test need be made only once; a compound could be stored immediately in TEMPARRAY0 when it contains only one element.

The specifications of a mask (what to look for in a set of relative locations) could be handled with 2-dimensional masks of the sort used in PICAP. This would be especially convenient if implemented as part of an interactive prompting routine. Masks should be declarable, e.g. with ||mask := {mask}; so that the programmer could simply name them, and also could perform operations on them, to transform them.

In addition, the format for declaring array-types and shrink-convergence can be regularized, generalized, and brought closer to standard PASCAL format. One example of a promising version would replace the declarations:

```
||procedure {procedurename};
  :
||dim & emerge from [..15,0..32] to [2,2];
  by
||procedure {name}(dim 0..15,0..15 : boolean; shrink 2,2);
```

(Now "shrink" would be only one of a number of conversion operations the programmer might designate are to be effected, along with, for example, "rotate" or "invert". The programmer should probably also be given the facility to write procedures that compute still other operations, and call them along with the built-in operations.)

PascalPL could also get the dimensions of arrays from the programmer's declarations, so that the procedure could be declared even more simply, e.g.:

|procedure {procedurename} (shrink 3,4);

Alternately, shrink (and other operations) could be handled as simple assignment statements (either with or without "||").

The dimensions might be used to specify sub-arrays, and then the programmer be given facilities for coding different sequences of instructions to be executed over different sub-arrays. This begins to give facilities for programming MIMD (Multiple-Instruction-Multiple-Data Stream) as well as SIMD arrays. In the extreme, the programmer could specify a sub-array of 1 cell, and different instructions for each such (1-cell) sub-array. It is not clear whether this would violate the parallel array structure (such programs certainly could not be executed except with enormous inefficiencies on actual arrays of physically parallel processors). But is seems an interesting step toward a language for more general networks, where the lock-step parallel processes of the array are relaxed. And it makes clear that it is not the MIMD vs. SIMD distinction that is important for efficiency in executing programs (as opposed to simplicity in building one program controller to drive all processors) but rather the need to keep all processors working for (close to) the same amount of time (which is guaranteed when all processors execute the same sequence of instructions).

A parallel version of the standard PASCAL Case statement would probably be desirable, to handle situations where a sequence of embedded "||if..||then.." statements would otherwise be needed. A slight extension should also be made, to allow inequalities over integer values. This would, for example, handle situations where one set of actions should be taken if a result of a compounding operation exceeded some threshold, while different sets of actions should be taken for intervals to successively smaller next threshold(s). Therefore the Case statement should accept inequalities as well as PASCAL character ("char") symbols, e.g.:

```
|case threshold of
  >57 : modify(hor+23,vert-3,tree+5);
  >21 : modify(noise+7,hor-3);
  >3  : modify(noise+11,vert-7);
  end {case}
```

This would be another step toward handling more general MIMD processes. A ||repeat...||until...; and a ||while...||do...; should be introduced, along with ||if...||then...||else...; statements that applied arbitrary blocks of code to the individual cells, rather than the relatively standard and therefore close-to-equal-in-time modifiers presently programmed.

6. DISCUSSION

Again, these changes would move the language away from a language for physically parallel arrays (and may well violate the spirit of parallel processing on these arrays) and toward much more general networks of processors. But if the programmer took care to keep all eventualities relatively close in the time they needed for execution, and appropriate hardware were available, this might be a good procedure for getting efficient programs. For the language would handle the allocation of processors and message passing automatically and efficiently. And the burden on the programmer would be relatively small—to formulate the program so that it is executed through a parallel-window-like set of processes of a size commensurate with the size of the hardware–parallel system (which might be an array, or some other appropriate network).

Should a language for parallel arrays and, more generally, for networks, remain as an extension embedded in PASCAL? The present and projected extensions appear to cohabit rather congenially within PASCAL. PascalPL seems surprisingly simple in the extensions that were needed, and it makes use of standard PASCAL in many ways. A programmer should be able to code in this mixture with little interference between the two systems. And PASCAL is useful when the programmer codes at least some processes for a serial computer (as she/he certainly will, at least until all the very difficult problems of parallel arrays and networks are solved).

But it seems likely that the parallel aspects of computing are of overriding importance, that we are entering a completely new era of parallel networks of computers. We will need to develop completely new parallel algorithms and programs; we will find that whole new types of approaches to problems, and of problems themselves, are now amenable to attack, and invite attack. At some point, possibly quite soon, entirely new languages will be called for. But it seems best to move toward new languages by first extending those that exist, and also trying to combine the features of different types of languages that appear to be relevant (e.g. APL, array languages like PICAP's PPL and CLIP's CAP4 [28], and multi-processor languages like Concurrent PASCAL and Modula). These endeavours, along with the continuing design of and experience with arrays and networks, should give us the understanding needed to develop entirely new and more appropriate languages.

REFERENCES

1] Anon (1979). Advertising brochure and personal communication. Goodyear-Aerospace, Akron, Ohio.

[2] Batcher, K. E. (1974). "STARAN parallel processor system hardware". Proc. AFIPS Nat. Comp. Conf., **43**, 405–410.

[3] Brinch Hansen, P. (1973). *Operating system principles*. Prentice-Hall, Englewood Cliffs.

[4] Douglass, R. J. (1979). "Extensions to PASCAL for parallel image processing". Paper presented at the Workshop on High Level Languages for Image Processing, Windsor, England.

[5] Duff, M. J. B. (1976). "CLIP4: a large scale integrated circuit array parallel processor". Proc. 3rd IJCPR, Coronado, USA, pp. 728–733.

[6] Duff, M. J. B. (1978). "Review of the CLIP image processing system". Proc. Nat. Comp. Conf., Anaheim, USA, pp. 1055–1060.

[7] Duff, M. J. B. (1979). "Final report on the Workshop on high level languages for image processing". University College London.

[8] Flanders, P. M., Hunt, D. J., Reddaway, S. F. and Parkinson, D. (1977). "Efficient high speed computing with the Distributed Array Processor". In *High speed computer and algorithm organisation*, pp. 113–128. Kuck, Lawrie and Sameh (eds). Academic Press, London and New York.

[9] Fung. L., "A massively parallel processing computer". In *High speed computer and algorithm organisation*. Kuck, Lawrie and Sameh (eds). Academic Press, New York and London.

[10] Gundmundsson, B. (1979). "An interactive high level language system for picture processing", Windsor Workshop.

[11] Ison, R. (1977). Unpublished paper on extensions to PASCAL for parallel networks, Univ. of Virginia.

[12] Iverson, K. E. (1962). **A Programming Language**. Wiley, New York.

[13] Jensen, K. and Wirth, N. (1976). *PASCAL user manual and report*, 2nd edn. Springer-Verlag, Berlin.

[14] Kruse, B. (1976). "The PICAP picture processing laboratory". Proc. 3rd IJCPR, Coronado, USA, pp. 875–881.

[15] Kruse, B. (1978). "Experience with a picture processor in pattern recognition processing". Proc. Nat. Comp. Conf., Anaheim, USA.

[16] Lawrie, D. H., Layman, T., Baer, D. and Randal, J. M. (1975). "GLYPNIR—a programming language for ILLIAC IV". *Comm. ACM* **18**, 157–164.

[17] Levialdi, S., Maggiolo-Schettini, A., Napoli, M. and Uccella, G. (1981). "PIXAL: a high level language for image processing". In *Real-time parallel computing*. Onoe, M. and Preston, K., Jr. (eds). Plenum Publishing Co., New York.

[18] Perrot, R. and Stevenson, D. (1978). "ACTUS—a language for SIMD architectures". Proc. LASL Workshop on Vector and Parallel Processors, Los Alamos, pp. 212–218.

[19] Preston, K., Jr. (1979). "Image manipulative languages: a preliminary survey". Windsor Workshop.

[20] Reddaway, S. F. (1978). "DAP—a flexible number cruncher". Proc. LASL Workshop on Vector and Parallel Processors, Los Alamos, pp. 233–234.

[21] Reeves, A. P. (1979). "An array processing system with a FORTAN-based realization". *Comp. Graph. and Image Proc.* **9**, 267–281.

[22] Schmitt, L. (1979). Unpublished paper on parallel languages for general networks, Univ. of Wisconsin.

3] Sternberg, S. R. "Cytocomputer real-time pattern recognition". Paper presented at the 8th Patt. Recog. Symp., Nat. Bur. Stand.

4] Travis, L., Honda, M., Le Blanc, R. and Zeigler, S. (1977). "Design rationale for TELOS, a PASCAL based AI language". *ACM SIGPLAN*, **12**, 67–76.

5] Uhr, L. (1979). "A language for programming scene description and pattern recognition systems on a parallel array computer". Windsor Workshop. (Also Tech. Rep. 354, Comp. Sci. Dept., Univ. of Wisconsin.)

6] Wirth, N. (1971). "Design of a PASCAL compiler". *Software—Practice and Experience*, **1**, 309–333.

7] Wirth, N. (1977). "Toward a discipline of real-time programming". *Comm. ACM*, **20**, 577–583.

8] Wood, A. (1977). *CAP4 Programmer's Manual*. Image Processing Group, University College London.

APPENDIX A

PRESENTLY IMPLEMENTED OPTIONS TO PASCALPL CONSTRUCTS

A number of optional fortmats are presently allowed, so that, in this experimental version, they can be compared in terms of convenience and preference. The spirit of PASCAL suggests that optional forms not be given users. But those that prove to be useful, and do not lead to more programming errors, might be worth keeping.

Present options include:

(1) Any number of vertical bars can identify a parallel construct, e.g. "|set" "|||||if..|then..|||else" (two bars, e.g. ||set, is recommended, since this is a standard symbol for parallel).

(2) "||let" can be used instead of "||set".

(3) "&emerge" or "bmerge" or "&" or "b" can be used to designate boolean arrays; to designate integer arrays.

The construct [xmin..xmax,ymin..ymax : boolean] can be used to declare a boolean (or ..: integer] for integer) array within the dimensions, as discussed above.

(4) Because (when they are implemented) a period must precede a name of an array in an array of records, the options are given to use ",", or "+" or "   " to indicate that ordinary arrays are used.

(5) ||dim & [0..7,0..7]; can be used instead of ||dim &from [0..7,0..7]; and, if the desired dimensions have previously been specified and the default condition of integer arrays is desired, then it is sufficient to write:

||dim; (which is equivalent to ||dim {integer} [{current dimensions}];).

(6) ||border := {bordertype} is not needed;
   border := {bordertype} will suffice.

But the former seems good practice, since border must be designated for the parallel array operations.

(7) Similarly, ||end; is not needed: end; will suffice.

(8) A sequence of arraynames for assignment or modification (in conditional statements) may be separated by mixtures of commas, end-brackets, or spaces, e.g.:

||set name1, name2, name3 :=
||set name1] name2, name 3 :=
||set name1, name2 name3] name4 :=

(9) If no arithmetic operator follows a name to be modified in a conditional, that array is assigned the value of the expression preceding ||then (and is therefore handled exactly as it would be by a ||set assignment statement).

(10) In assignment statements, names can be followed by arithmetic

operators and integers. But these will be taken to specify modifications to the assigned value, rather, than, as in the conditional statements, as modifications to the presently stored values.

(11) The ‖read(. .); and ‖write(. .); constructs can specify sub-arrays, e.g., ‖read(image[0 . .3,2 . .7]);. But this must be done carefully in the present implementation, since these become the new array dimensions, and must lie within the previous dimensions.

(12) ‖vari {integer variables} : integer; and
    ‖varb {boolean variables} : boolean;
can be used as an alternative way of declaring array type. (This should be more useful when the program is extended to handle automatically the finding and using of the array's type in determining what type of operations to perform.)

APPENDIX B

SYNTAX DIAGRAMS FOR THE PASCALPL EXTENSIONS TO PASCAL

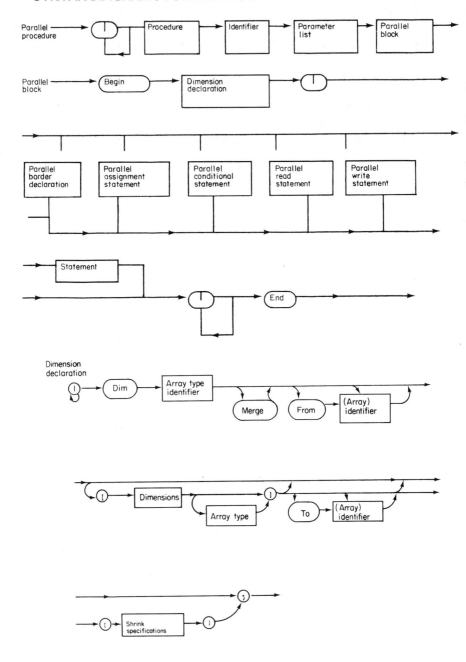

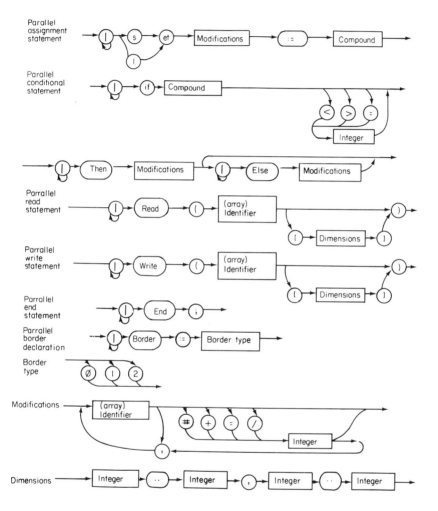

Parallel assignment statement

Parallel conditional statement

Parallel read statement

Parallel write statement

Parallel end statement

Parallel border declaration

Border type

Modifications

Dimensions

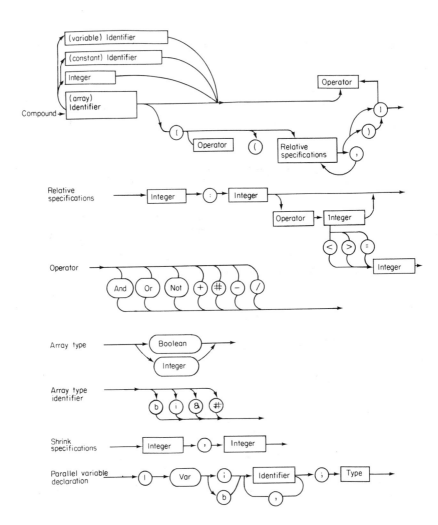

EXAMPLES OF PROGRAM AND CODE TRANSLATED FROM PASCALPL TO PASCAL

The PascalPL program input to the PascalPL preprocessor

```
program simple(input,output);
    {the programmer must declare the array data structures used}
    const
      xmin = 0; xmax = 7; ymin = 0; ymax = 7;
    type
      xindex = xmin . .xmax;
      yindex = ymin . .ymax;
      arraytype = array [xindex,yindex] of integer;
    var
      arrayx : xmin . .xmax;
      arrayy : ymin . .ymax;
      image, negative, edgedimage : arraytype;

    procedure sayhello;
    begin
      writeln('hello');
    end;

    ||procedure demonstrate;
    begin
    ||dim      [0 . .2,0 . .2];
      ||read(image);
      ||write (image);
        ||read(negative);
      writeln('the image and the negative image have been input');
      ||set edgedimage := image − negative;
      ||write(edgedimage);
      ||set image, negative := 0;
    end;

    begin {program}
      sayhello;
      demonstrate;
    end.
```

The PASCAL program output for the simple program above

{****STARTING READIN****}
{****YOU MAY ASSIGN border := 0 (or 1 or 2)****}
{****DEFAULT border = 2 (for self − i.e., the nearest inner border)****}
{****DEFAULT arraydimensions are 0 ..7, 0 ..7****}
{****DEFAULT xshrink, yshrink are 1 and 1 (no shrink)****}

(*INPUT1=program simple(input,output);*)
program simple(input,output);

(*INPUT2= {the programmer must declare the array data structures used} *)
  {the programmer must declare the array data structures used}

(*INPUT3=const *)
  const

(*INPUT4= xmin = 0; xmax = 7; ymin = 0; ymax = 7; *)
  xmin = 0; xmax = 7; ymin = 0; ymax = 7;

(*INPUT5= type*)
  type

(*INPUT6= xindex = xmin ..xmax; *)
  xindex = xmin..xmax; *)

(*INPUT7= yindex = ymin ..ymax; *)
  yindex = ymin ..ymax;

(*INPUT8= arraytype = array [xindex,yindex] of integer; *)
  arraytype = array [xindex,yindex] of integer;

(*INPUT9= var *)
  var

(*INPUT10= arrayx : xmin ..xmax; *)
  arrayx : xmin ..xmax;

(*INPUT11= arrayy : ymin ..ymax; *)
  arrayy : ymin ..ymax;

(*INPUT12= image, negative, edgedimage : arraytype; *)
  image, negative, edgedimage : arraytype;

(*INPUT13= *)

(*INPUT14= procedure sayhello; *)
procedure sayhello;

(*INPUT15= begin *)
begin

(*INPUT16= writeln('hello'); *)
  writeln('hello');

(*INPUT17= end; *)
end;

(*INPUT18= *)

(*INPUT19= ‖procedure demonstrate; *)
procedure plparpll;
        forward;

                {**   1‖PARALLEL BLOCK starts.**}
procedure demonstrate;

(*INPUT20= begin *)
begin

(*INPUT21= ‖dim   [0..2,0..2]; *)
    plparpll;
{****   DIMENSIONS ARE=   0   2   0   2****}
end;
                {***starting a block of PARALLEL CODE!!**}
procedure plparpll;
const
    xmin = 0;
    xmax = 2;
    ymin = 0;
    ymax = 2;
type
  xindex = xmin..xmax;
  yindex = ymin..ymax;
    temparraytype = array [xindex,yindex] of integer;
var
  beyondborder : boolean;

```
    partineq : char;
    valueread, border, tempstore, partthresh, arrayxloc,arrayyloc,
    arrayx : xmin..xmax;                        xshrink,yshrink : integer;
    arrayy : ymin..ymax;
    TEMPARRAY0, TEMPARRAY1, TEMPARRAY2 : temparraytype;
begin
    border := 2;
{****SET NO SHRINK, so SHRINK = 1,1****}
    xshrink := 1;
    yshrink := 1;

(*INPUT22= ||read(image); *)
    writeln("TYPE IN INTEGERS FOR image[0..2, 0..2]");
      for arrayx :=   0 to   2 do
        begin
          for arrayy :=   0 to   2 do
            begin
              read(valueread);
              image[arrayx,arrayy] := valueread;
              writeln("[",arrayx," , ",arrayy,"] := ",valueread);
            end;
        end;

(*INPUT23= ||write(image); *)
    writeln("ARRAY = image contains:");
      for arrayx :=   0 to   2 do
        begin
          for arrayy :=   0 to   2 do
            begin
              write(image[arrayx,arrayy]:4);
            end;
          writeln(";");
        end;
      writeln("END OF ARRAY .");

(*INPUT24= ||read(negative); *)
      writeln("TYPE IN INTEGERS FOR negative[0..2, 0..2]");
      for arrayx :=   0 to   2 do
        begin
          for arrayy :=   0 to   2 do
            begin
              read(valueread);
```

```
                negative[arrayx,arrayy] := valueread;
                writeln("[",arrayx," , ",arrayy,"] := ",valueread);
          end;
     end;

(*INPUT25 = writeln('the image and the negative image have been input'); *)
     writeln('the image and the negative image have been input');

(*INPUT26 = ||set edgedimage := image – negative; *)
     for arrayx :=  0 to  2 do
          begin
             for arrayy :=  0 to  2 do
                begin
                   {(*NO REL LOCS, SO NO BORDER TEST NEEDED*)}
                   arrayxloc := arrayx;
                   arrayyloc := arrayy;
                   if beyondborder = true   then
                         TEMPARRAY1[arrayx,arrayy] := 0
                   else
                         begin
                            tempstore := image[arrayxloc,arrayyloc];
                            TEMPARRAY1[arrayx,arrayy] := tempstore;
                         end;
                   TEMPARRAY0[arrayx,arrayy] :=
                                      TEMPARRAY1[arrayx,arrayy];
                   {(*NO REL LOCS, SO NO BORDER TEST NEEDED*)}
                   arrayxloc := arrayx;
                   arrayyloc := arrayy;
                   if beyondborder = true   then
                         TEMPARRAY1[arrayx,arrayy] := 0
                   else
                         begin
                            tempstore := negative[arrayxloc,arrayyloc];
                            TEMPARRAY1[arrayx,arrayy] := tempstore;
                         end;
                   TEMPARRAY0[arrayx,arrayy] :=
                                      TEMPARRAY0[arrayx,arrayy]
                                    – TEMPARRAY1[arrayx,arrayy];
                end;
          end;
     for arrayx :=  0 to  2 do
          begin
```

```
      for arrayy := 0 to   2 do
         begin
         edgedimage[arrayx div xshrink,arrayy div yshrink] :=
                                      TEMPARRAY0[arrayx,arrayy];
         end;
      end;

(*INPUT27= ||write(edgedimage); *)
  writeln("ARRAY = edgedimage contains:");
     for arrayx := 0 to   2 do
        begin
        for arrayy := 0 to   2 do
           begin
              write(edgedimage[arrayx,arrayy]:4);
           end;
        writeln(";");
        end;
     writeln("END OF ARRAY .");

(*INPUT28= ||set image, negative := 0; *)
     for arrayx := 0 to   2 do
        begin
        for arrayy := 0 to   2 do
           begin
              TEMPARRAY0[arrayx,arrayy] := 0;
           end;
        end;
     for arrayx := 0 to   2 do
        begin
        for arrayy := 0 to   2 do
           begin
           image[arrayx div xshrink,arrayy div yshrink] :=
                                      TEMPARRAY0[arrayx,arrayy];
           negative[arrayx div xshrink,arrayy div yshrink] :=
                                      TEMPARRAY0[arrayx,arrayy];
           end;
        end;

(*INPUT29= ||end; *)
end; {END OF PARALLEL BLOCK= 1}

(*INPUT30= *)
```

(*INPUT31= begin {program} *)
begin {program}

(*INPUT32= sayhello; *)
 sayhello;

(*INPUT33= demonstrate; *)
 demonstrate;

(*INPUT34= end . *)
end .

(*INPUT35= *)

(*INPUT36= *)
{****READIN HAS FINISHED.****}

EXAMPLES OF ||SET STATEMENTS TRANSLATED FROM PASCALPL TO PASCAL

PASCAL code follows for the following two PascalPL statements:
 ||set cross := vert[+(0:0,0:2,0:−2)] * hor[+(0:0,2:0−2:0)];
 ||set featurei, featurej, labelk
     := featurem[*(4:−2>5,−5:7*3>14,0:0*21>112)];

(*INPUTx= ||set cross := vert[+(0:0,0:2,0:−2) * hor[+(0:0,2:0,−2:0)]; *)
{****ASSIGNMENT STATEMENT HANDLED HERE.****}
  for arrayx :=   0 to   2 do
    begin
      for arrayy :=   0 to   2 do
        begin
          arrayxloc := arrayx + 0;
          arrayyloc := arrayy + 0;
                    {(**KEEPS IT IN BOUNDS!**)}
          beyondborder := false;
            if arrayxloc <   0   then beyondborder := true;
            if arrayxloc >   2   then beyondborder := true;
            if arrayyloc <   0   then beyondborder := true;
            if arrayyloc >   2   then beyondborder := true;
          if beyondborder = true   then
            TEMPARRAY1[arrayx,arrayy] := 0

```
    else
      begin
        tempstore := vert[arrayxloc,arrayyloc]  *  1;
        TEMPARRAY1[arrayx,arrayy] := tempstore;
      end;
    arrayxloc := arrayx + 0;
    arrayyloc := arrayy + 2;
                {(**KEEPS IT IN BOUNDS!**)}
    beyondborder := false;
      if arrayxloc <  0  then beyondborder := true;
      if arrayxloc >  2  then beyondborder := true;
      if arrayyloc <  0  then beyondborder := true;
      if arrayyloc >  2  then beyondborder := true;
    if beyondborder = true   then
      TEMPARRAY2[arrayx,arrayy] :=
                              0 + TEMPARRAY1[arrayx,arrayy]
    else
      begin
        tempstore := vert[arrayxloc,arrayyloc]  *  1;
        TEMPARRAY2[arrayx,arrayy] :=
                              TEMPARRAY1[arrayx,arrayy]
                                        + tempstore;
        if border < 2 then
          if TEMPARRAY2[arrayx,arrayy] > 1 then
            TEMPARRAY2[arrayx,arrayy] := 0;
      end;
    arrayxloc := arrayx + 0;
    arrayyloc := array + -2;
                {(**KEEPS IT IN BOUNDS!**)}
    beyondborder := false;
      if arrayxloc <  0  then beyondborder := true;
      if arrayxloc >  2  then beyondborder := true;
      if arrayyloc <  0  then beyondborder := true;
      if arrayyloc >  2  then beyondborder := true;
    if beyondborder = true   then
      TEMPARRAY1[arrayx,arrayy] :=
                              0 + TEMPARRAY2[arrayx,arrayy]
    else
      begin
        tempstore := vert[arrayxloc,arrayyloc] * 1;
        TEMPARRAY1[arrayx,arrayy] :=
                              TEMPARRAY2[arrayx,arrayy]
                                        + tempstore;
```

```
              if border < 2 then
                 if TEMPARRAY1[arrayx,arrayy] > 1 then
                                TEMPARRAY1[arrayx,arrayy] := 0;
              end;
      TEMPARRAY0[arrayx,arrayy] := TEMPARRAY1[arrayx,arrayy];
arrayxloc := arrayx + 0;
arrayyloc := arrayy +0;
                            {(**KEEPS IT IN BOUNDS!**)}
        beyondborder := false;
           if arrayxloc <  0   then beyondborder := true;
           if arrayxloc >  2   then beyondborder := true;
           if arrayyloc <  0   then beyondborder := true;
           if arrayyloc >  2   then beyondborder := true;
        if beyondborder = true then
                TEMPARRAY1[arrayx,arrayy] := 0
           else
              begin
                 tempstore := hor[arrayxloc,arrayyloc] * 1;
                 TEMPARRAY1[arrayx,arrayy] := tempstore;
              end;
arrayxloc := arrayx + 2;
arrayyloc := arrayy + 0;
                            {(**KEEPS IT IN BOUNDS!**)}
        beyondborder := false;
           if arrayxloc <  0   then beyondborder := true;
           if arrayxloc >  2   then beyondborder := true;
           if arrayyloc <  0   then beyondborder := true;
           if arrayyloc >  2   then beyondborder := true;
        if beyondborder = true   then
                TEMPARRAY2[arrayx,arrayy] :=
                             0 + TEMPARRAY1[arrayx,arrayy]
           else
              begin
                 tempstore := hor[arrayxloc,arrayyloc] * 1;
                 TEMPARRAY2[arrayx,arrayy] :=
                                    TEMPARRAY1[arrayx,arrayy]
                                          + tempstore;
              if border < 2 then
                 if TEMPARRAY2[arrayx,arrayy]   > 1   then
                    TEMPARRAY2[arrayx,arrayy] := 0;
              end;
arrayxloc := arrayx + −2;
```

```
            arrayyloc := arrayy + 0;
                      {(**KEEPS IT IN BOUNDS!**)}
            beyondborder := false;
              if arrayxloc <  0   then beyondborder := true;
              if arrayxloc >  2   then beyondborder := true;
              if arrayyloc <  0   then beyondborder := true;
              if arrayyloc >  2   then beyondborder := true;
            if beyondborder = true   then
              TEMPARRAY1[arrayx,arrayy] :=
                              0 + TEMPARRAY2[arrayx,arrayy]
            else
              begin
                tempstore := hor[arrayxloc,arrayyloc] * 1;
                TEMPARRAY1[arrayx,arrayy] :=
                              TEMPARRAY2[arrayx,arrayy]
                                          + tempstore;
                if border <  2   then
                  if TEMPARRAY1[arrayx,arrayy]  > 1   then
                    TEMPARRAY1[arrayx,arrayy] := 0;
              end;
            TEMPARRAY0[arrayx,arrayy] :=
                              TEMPARRAY0[arrayx,arrayy]
                              * TEMPARRAY1[arrayx,arrayy];
          end;
        end;
      for arrayx :=  0 to  2 do
        begin
          for arrayy :=  0 to  2 do
            begin
            cross[arrayx div xshrink,arrayy div yshrink] :=
                              TEMPARRAY0[arrayx,arrayy];
          end;
        end;

(*INPUTy1= ||set featurei, featurej, lablelk *)
{****ASSIGNMENT STATEMENT HANDLED HERE .****}

(*INPUTy2=   := featurem[*(4:−3*2>5,−5:7*3>14,0:0*21>112)]; *)
    for arrayx :=  0 to  2 do
      begin
        for arrayy :=  0 to  2 do
          begin
```

```
arrayxloc := arrayx + 4;
arrayyloc := arrayy + −3;
                {(**KEEPS IT IN BOUNDS!**)3;
    beyondborder := false;
    if arrayxloc <   0   then beyondborder := true;
    if arrayxloc >   2   then beyondborder := true;
    if arrayyloc <   0   then beyondborder := true;
    if arrayyloc >   2   then beyondborder := true;
    if beyondborder = true   then
       TEMPARRAY1[arrayx,arrayy] := 0
    else
       begin
          tempstore := featurem[arrayxloc,arrayyloc] * 2;
             if not (tempstore > 5) then
                tempstore := 0;
             TEMPARRAY1[arrayx,arrayy] := tempstore;
       end;
arrayxloc := arrayx + −5;
arrayyloc := arrayy + 7;
                {(**KEEPS IT IN BOUNDS!**)}
    beyondborder := false;
    if arrayxloc <   0   then beyondborder := true;
    if arrayxloc >   2   then beyondborder := true;
    if arrayyloc <   0   then beyondborder := true;
    if arrayyloc >   2   then beyondborder := true;
    if beyondborder = true   then
       TEMPARRAY2[arrayx,arrayy] :=
       0 * TEMPARRAY1[arrayx,arrayy]
    else
       begin
          tempstore := featurem[arrayxloc,arrayyloc] * 3;
             if not (tempstore > 14) then
                tempstore := 0;
             TEMPARRAY2[arrayx,arrayy] :=
                               TEMPARRAY1[arrayx,arrayy]
                                              * tempstore;
          if border < 2   then
             if TEMPARRAY2[arrayx,arrayy] > 1   then
                TEMPARRAY2[arrayx,arrayy] := 0;
       end;
arrayxloc := arrayx + 0;
arrayyloc := arrayy + 0;
```

```
                    {(**KEEPS IT IN BOUNDS!**)}
          beyondborder := false;
            if arrayxloc <  0  then beyondborder := true;
            if arrayxloc >  2  then beyondborder := true;
            if arrayyloc <  0  then beyondborder := true;
            if arrayyloc >  2  then beyondborder := true;
            if beyondborder = true  then
            TEMPARRAY1[arrayx,arrayy] :=
                              0 * TEMPARRAY2[arrayx,arrayy]
        else
            begin
              tempstore := featurem[arrayxloc,arrayyloc] * 21;
                if not (tempstore >  112)  then
                  tempstore := 0;
                  TEMPARRAY1[arrayx,arrayy] :=
                                  TEMPARRAY2[arrayx,arrayy]
                                                  * tempstore;
                if border <  2  then
                if TEMPARRAY1[arrayx,arrayy] >  1  then
                    TEMPARRAY1[arrayx,arrayy] := 0;
            end;
            TEMPARRAY0[arrayx,arrayy] :=
            TEMPARRAY1[arrayx,arrayy];
        end;
      end;
    for arrayx :=  0 to  2 do
      begin
        for arrayy :=  0 to  2 do
          begin
            featurei[arrayx div xshrink,arrayy div yshrink] :=
                                  TEMPARRAY0[arrayx,arrayy];
            featurej[arrayx div xshrink,arrayy div yshrink] :=
                                  TEMPARRAY0[arrayx,arrayy];
            labelk[arrayx div xshrink,arrayy div yshrink] :=
                                  TEMPARRAY0[arrayx,arrayy];
          end;
      end;
```

AN EXAMPLE OF PASCALPL OUTPUT FOR A SIMPLE CONDITIONAL STATEMENT

The following shows the PascalPL conditional statement (labelled INPUTz), and then the PASCAL code output by PascalPL, for:

```
||if featurei[+(0:1,0:−1)] * featurej[+(1:0,−1:0)] > 11
||then    labeli+19, labelj*2, labelk−33
||else    labell*2, labelm+27;
```

```
(*INPUTz1 = ||if featurei[+(0:1,0:−1)) * featurej[+(1:0,−1:0)] > 11*)
        {IF STATEMENT BEING HANDLED HERE.}
    for arrayx :=   0 to   2 do
      begin
        for arrayy :=   0 to   2 do
          begin
            arrayxloc := arrayx + 0;
            arrayyloc := arrayy + 1;
                        {(**KEEPS IT IN BOUNDS!**)}
            beyondborder := false;
              if arrayxloc <   0   then beyondborder := true;
              if arrayxloc >   2   then beyondborder := true;
              if arrayyloc <   0   then beyondborder := true;
              if arrayyloc >   2   then beyondborder := true;
            if beyondborder = true   then
              TEMPARRAY1[arrayx,arrayy] := 0
            else
              begin
                tempstore := featurei[arrayxloc,arrayyloc] * 1;
                TEMPARRAY1[arrayx,arrayy] := tempstore;
              end;
            arrayxloc := arrayx +0;
            arrayyloc := arrayy + −1;
                        {(**KEEPS IT IN BOUNDS!**)}
            beyondborder := false;
              if arrayxloc <   0   then beyondborder := true;
              if arrayxloc >   2   then beyondborder := true;
              if arrayyloc <   0   then beyondborder := true;
              if arrayyloc >   2   then beyondborder := true;
            if beyondborder = true   then
              TEMPARRAY2[arrayx,arrayy] :=
                        0 + TEMPARRAY1[arrayx,arrayy]
```

```
else
  begin
    tempstore := featurei[arrayxloc,arrayyloc] * 1;
    TEMPARRAY2[arrayx,arrayy] :=
                           TEMPARRAY1[arrayx,arrayy]
                                          + tempstore;
    if border <  2  then
      if TEMPARRAY2[arrayx,arrayy] >  1   then
        TEMPARRAY2[arrayx,arrayy] := 0;
  end;
  TEMPARRAY0[arrayx,arrayy] :=
                           TEMPARRAY2[arrayx,arrayy];
arrayxloc := arrayx + 1;
arrayyloc := arrayy + 0;
          {(**KEEPS IT IN BOUNDS!**)}
  beyondborder := false;
  if arrayxloc <  0   then beyondborder := true;
  if arrayxloc >  2   then beyondborder := true;
  if arrayyloc <  0   then beyondborder := true;
  if arrayyloc >  2   then beyondborder := true;
if beyondborder = true   then
  TEMPARRAY1[arrayx,arrayy] := 0
else
  begin
    tempstore := featurej[arrayxloc,arrayyloc] * 1;
    TEMPARRAY1[arrayx,arrayy] := tempstore;
  end;
arrayxloc := arrayx + -1;
arrayyloc := arrayy + 0;
          {(**KEEPS IT IN BOUNDS!**)}
  beyondborder := false;
  if arrayxloc <  0   then beyondborder := true;
  if arrayxloc >  2   then beyondborder := true;
  if arrayyloc <  0   then beyondborder := true;
  if arrayyloc >  2   then beyondborder := true;
if beyondborder = true   then
  TEMPARRAY2[arrayx,arrayy] :=
                      0 + TEMPARRAY1[arrayx,arrayy]
else
  begin
    tempstore := featurej[arrayxloc,arrayyloc] * 1;
    TEMPARRAY2[arrayx,arrayy] :=
```

```
                                    TEMPARRAY1[arrayx,arrayy]
                                         + tempstore;
              if border <   2   then
                if TEMPARRAY2[arrayx,arrayy] > 1   then
                   TEMPARRAY2[arrayx,arrayy] := 0;
              end;
              TEMPARRAY0[arrayx,arrayy] :=
                                    TEMPARRAY0[arrayx,arrayy]
                                    * TEMPARRAY2[arrayx,arrayy];
          end;
       lxend;

(*INPUTz2=    ||then labeli+9, labelj*2, labelk-33 *)
   for arrayx :=   0 to   2 do
      begin
        for array :=   0 to   2 do
          begin
                (**modify if found greaterthan value**)
            if (TEMPARRAY0[arrayx,arrayy] > 11)   then
              labeli[arrayx div xshrink,arrayy div yshrink] :=
                      labeli[arrayx div xshrink,arrayy div yshrink] + 19;
                (**modify if found greaterthan value**)
            if (TEMPARRAY0[arrayx,arrayy] > 11)   then
              labelj[arrayx div xshrink,arrayy div yshrink] :=
                      labelj[arrayx div xshrink,arrayy div yshrink] * 2;

(*INPUTz3=    ||else labell*2, labelm+27; *)
                (**modify if found greaterthan value**)
            if (TEMPARRAY0[arrayx,arrayy] > 11)   then
              labelk[arrayx div xshrink,arrayy div yshrink] :=
                      labelk[arrayx div xshrink,arrayy div yshrink] - 33;

            else
                (**else do the following:**)
              labell[arrayx div xshrink,arrayy div yshrink] :=
                      labell[arrayx div xshrink,arrayy div yshrink] * 2;
                (**else do the following:**)
              labelm[arrayx div xshrink,arrayy div yshrink] :=
                      labelm[arrayx div xshrink,arrayy div yshrink] + 27;
          end;
      end;
```

Chapter Six

On the Design and Implementation of PIXAL, a Language for Image Processing

S. Levialdi, A. Maggiolo-Schettini, M. Napoli,
G. Tortora and G. Uccella

1. INTRODUCTION

In order to increase the overall efficiency in numerical computation, for many years different strategies have been suggested; i.e. whole word access, multi-programming, pipelining, and more recently, the development of various multiprocessor architectures. Nevertheless, the real gain in the execution time allowed by parallel architectures is highly dependent on the specific algorithm to be implemented, hence the efforts to make parallel algorithms which are currently used in numerical computation. A field which conveniently exploits parallel architectures is image processing, since its data structures consist of a large number of elements (pixels) which may be processed independently. Generally the scanning of an image and its sequential processing are only due to the sequential nature of the computers that have been used so far. In many cases the new value of each picture element is obtained in terms of the values of its local neighbours and this holds for all the pixels of the image independently so that the new image may be obtained in one single computational step.

For many years researchers in image processing have tried to design software tools which make better use of the capabilities of existing computers. As a result of this effort (a) special subroutines were written that could be called by a FORTRAN program [see 7, 11]; (b) a LISP compiler has been extended so as to improve the display of image properties and image operations [see 5]; (c) a number of assembly languages in connection with image processing computers have also been designed and used [see 2, 6].

A number of projects are under way, like the design of a fast interactive man–machine communication for specifying image transformations and the definition of high-level languages for easily writing image processing algorithms. It has been our feeling that a language which allows the possibility of naturally expressing both global and parallel operations (without excluding the conventional sequential ones) and which also offers portability and transparency of structures could be extremely useful to the image processing community.

Along these lines we have formally defined an algorithmic high-level language with parallel facilities called PIXAL (PIXel manipulation Algorithmic Language) [see 8, 9].

The final goal of this project, which is currently developed in Naples, Pisa and Salerno, is to be able to program a machine of the SIMD kind (like, for instance, a CLIP machine [see 3]).

As a first step, a compiler for PIXAL is being written in the HP Assembly language of the HP 21MX available to all the groups involved in the project, with two possible options: HP object code and CAP4 object code for running on the CLIP4 machine.

The interest in running a PIXAL program on a sequential machine is due to the following facts: (a) it is useful to write programs implementing parallel algorithms in a parallel language; (b) it helps in assessing the speed-up gain that could be obtained on a parallel machine by means of conversion factors appended on each PIXAL construct; (c) when the CAP4 option is used, the program is directly transportable to the CLIP machine.

With the aim of having a high-level language for image processing with both sequential and parallel capabilities and also containing structures for numerical manipulation and computation we have decided not to define a language anew but to extend ALGOL 60. As a consequence of this approach, we have started to modify properly the available HP Algol Compiler. We will now describe the main new constructs which have been defined and characterize PIXAL. A syntactical definition of each construct will be given together with a description of the intended semantics and comments about the implementation on the HP 21MX.

The syntactic rules given in the following are intended to be added to those of the grammar of ALGOL 60, enriching the language with the new constructs. HP ALGOL is an LL(1) version of ALGOL 60 with some limitations on the recursive calls to procedures and on the dynamic allocation of blocks and with some non-standard ALGOL 60 constructs, for instance the **case**-construct [see 4]. (Note that the rules given here are not LL(1) but can be easily modified to obtain such form.)

2. PIXAL CONSTRUCTS

Besides the usual types of ALGOL 60 (real, integer and boolean), grey and binary types are also included for the representation of grey and binary picture elements respectively. Together with the possibility of defining arrays, which will be used for representing pictures, two other multidimensional structures, mask and frame, are introduced.

2.1 Syntax

<declaration>::= <mask declaration>|<frame declaration>|<edge
                                                   declaration>
<mask declaration>::= <type>**mask**<mask identifier><index list>|
                 <type>**mask**<mask identifier><index list>**of**
                                                   <value list>
<frame declaration>::= **frame**<frame identifier><index list>
<edge declaration>::=**edge-of**<array identifier>**is**<arithmetic expression>
<type>::= **binary|grey**
<mask identifier>::= <identifier>
<frame identifier>::= <identifier>
<index list>::= [<bound pair list>]|[<bound pair list>] **on**
                 (<arithmetic expression>{,<arithmetic expression>}]
<value list>::= (<extended arithmetic expression>{,<extended
                                        arithmetic expression>})
<extended arithmetic expression>::= <arithmetic expression>|<don't
                                                   care symbol>
<don't care symbol>::= ?
<procedure declaration>::=<type>**array**(<bound pair list>]**procedure**
                 <procedure heading><procedure body>
<specifier>::= <type>**mask|frame**

2.1.1 Semantics

Mask and frame declarations are used to declare certain identifiers to represent multidimensional structures. The list of indices gives the bounds in the different dimensions and, optionally, the co-ordinates of a special element which allows the positioning of masks and frames on an array element whenever the special element does not correspond to the geometrical centre. The list of values (among which a special "don't care" symbol, "?") defines the particular pattern of the mask to be compared with the environment of the element on which the mask is positioned.

The frame allows detection of a submatrix around the array element on which the frame is positioned (this will enable the definition of a neighbourhood upon which the parallel operation will be performed).

An edge declaration may be used for constraining the image data either to be embedded in a background (0-elements) or in any specific grey level whose value is provided. The default option automatically gives the first instance (this feature is generally implemented in the hardware of array processors [see 3]). Note that when neighbourhoods partially fall outside the array, the built-in functions use the edge declaration to ensure correct operation.

2.1.2 Implementation

The routine of the Algol 60 Compiler which analyses the type declaration of variable identifiers must be modified in order to consider binary and grey types. Variables declared as binary will require only one bit of memory; for those declared grey type, one byte will be needed (the maximum grey level value is $2^8 - 1 = 255$).

The routine which parses the array declaration builds an array template. The array template also contains the information regarding the edge. The proper field of the template is set to zero until the edge declaration is possibly found and set to the declared value.

The edge declaration may be changed on entering a block without changing the declaration of the array (only the information about the edge in the template of the array will be changed).

New routines parse the mask and frame declarations and build the corresponding templates. The mask template contains the dimensions of the mask, the co-ordinates of the special element, a pointer to the value list and the type of the values. The frame template contains only the dimensions of the frame and the co-ordinates of the special element.

Parallelism in PIXAL can be expressed in two different ways, both of which represent simultaneous processing of all the pixels but differ in the diameter size of the predicate which is being computed. In one case, the diameter is limited to a nearest neighbourhood (local operation) whilst in the second case, the diameter coincides with the whole array, therefore testing all the pixels for a given predicate (global operation).

The key control structure of PIXAL which allows natural expression of the parallel computations to be performed on an image is the **par-parend** construct.

2.2 Syntax

<parallel statement>::= **par**<special statement>**parend**
<special statement>::= any ALGOL 60 statement in which labels and **gotos**
are not allowed, subscripted variables do not occur
and assignments can only be made to array
identifiers.

2.2.1 Semantics

The special statement inside **par-parend** is intended to be performed simultaneously for every element of each array appearing within the special statement. As the special statement may also be a compound statement, an unlimited number of statements can occur inside **par-parend**.

2.2.2 Implementation

As we are implementing PIXAL on a sequential machine, the actual execution of a parallel statement must simulate the parallelism. A memory area must be allocated in order to contain as many arrays as appear on the left of the assignment statements inside the special statement. As an example, let us suppose that only one array A is involved and that the special statement S must be performed. B is the computed array for which an area is allocated; for every element A(i,j) of A the execution will be as follows:

(1) A(i,j) will be saved in a temporary location T;
(2) The statement S is executed;
(3) The updated value of A(i,j) is stored in B(i,j);
(4) The value of A(i,j) in T is restored in A(i,j).

Note that when a statement S is executed for i and j, every element of A can be tested but only A(i,j) is modified so that only one location for T is required. After the execution of the steps (1) to (4) for every element of A the resulting array can be found in B and hence it must be stored in A.

An optional assignment statement of PIXAL, of global nature, allows simultaneous comparison of the values of the elements of an array without specifying the indices.

2.3 Syntax

<assignment statement>::= <special assignment statement>|<array
identifier>:=<arithmetic expression>|
<procedure identifier>:=<special arithmetic

expression>|<mask identifier>:=<value list>
<special assignment statement<::= <array identifier>:=<special
arithmetic expression>
<special arithmetic expression>::= any arithmetic expression in which
array identifiers are allowed and
subscripted variables are forbidden

2.3.1 Semantics

When the right-hand side of the assignment is an arithmetic expression, its value will be computed and assigned to every element of the array on the left-hand side. When the right-hand side is a special arithmetic expression, its evaluation will produce an array of values which will be simultaneously assigned to the left-hand side array. (Note that assignments to arrays can also be performed under sequential control.)

The assignment of a list of values to a mask identifier means the assignment of the values of the list to the elements of the mask by row order.

2.3.2 Implementation

The implementation of the global assignment is straighforward and exploits the semantic routines in the compiler implementing the "for" statement.

As well as the tests performable within the **par-parend** constructs using a boolean expression to compute a local predicate, alternatively we may compute a global predicate on an array if the boolean expression is not contained within the **par-parend** construct.

2.4 Syntax

<boolean expression>::= <special boolean expression>
<special boolean expression>::= any boolean expression in which array
identifiers are allowed and subscripted
variables are forbidden

2.4.1 Semantics

The special boolean expression evaluates to "true" if the boolean property is satisfied by all the elements of the array.

2.4.2 Implementation

Notice that the special statements have a different implementation whether they appear inside a **par-parend** construct or not. The top-down analyser will ensure the correct code generation on the basis of the detection of a **par** delimiter.

Certain identifiers should be reserved for some useful functions in pattern recognition and image processing. The reserved list contains, among others,

sum(A)—for the function which gives the sum of all the elements of the array A.

Inside a parallel statement the following functions can be used:

compare(M,A)— for the boolean function which, given a mask M and an array A, is true if the submatrix of A with the same dimensions as M and centred over the current element of A (selected by the parallel control) is equal to M, false otherwise.

overlap(F,A)— for the function which, given a frame F and an array A, provides a new array with the same dimensions as F and whose elements are the same as those in the environment of the current element of A.

overweigh(M,A)— for the function which, given a mask M and an array A, provides a new array with the same dimensions as M and whose elements are the products between M and the corresponding elements around the current element of A.

3. EXAMPLES OF PIXAL PROGRAMS

Here now are two examples of PIXAL programs performing thinning and thresholding which illustrate the use of the above constructs showing the simplicity and shortness of such coding.

3.1 Example 1

A parallel thinning algorithm given in [1] and using eight 3 × 3 masks may be coded so that the deletability of an element is checked and performed within **par-parend** construct where the last instruction is also used as a stopping test in the iteration.

program thinning
begin

binary array A[1:256,1:256];
binary mask M[1:3, 1:3];
binary array V[1:256, 1:256];**integer** i;
V:=1
comment 1: a control array V is used to stop the iteration as soon as no new
elements are deleted. In a more recent version the possibility of
testing for "no new elements updated" should be available **endcom**.
while V≠0 **do**
 begin
 V:=0; **for** i:=1 **to** 8 **do**
    begin
    case i
    M:=(0,0,?,0,1,1,?,1,?);
    M:=(0,0,0,?,1,?,1,1,?);
    M:=(?,0,0,1,1,0,?,1,?);
    M:=(1,?,0,1,1,0,?,?,0);
    M:=(?,1,?,1,1,0,?,0,0);
    M:=(?,1,1,?,1,?,0,0,0);
    M:=(?,1,?,0,1,1,0,0,?);
    M:=(0,?,?,0,1,1,0,?,1);
    endcase;
    par if compare(M,A) **then begin** A:=0 V:=1 **end parend**
    end;
  end;
end;
comment 2: the values given for a mask are updated to cover all the eight
required masks **endcom**.

3.2 Example 2

An algorithm for object extraction [see 10] is coded in which the threshold is
obtained by analysing the histogram of the pixels which have a high gradient
value (exceeding a given threshold).
program threshold
begin
grey array A[1:128,1:128];
binary array BIN[1:128,1:128];
integer array HIST[0:63];
integer M1[1:1,1:2] **on** [1,1] **of** (1,−1);
integer M2[1:1,1:2] **on** [1,2] **of** (−1,1);
integer M3[1:2,1:1] **on** [1,1] **of** (1,−1);

integer M4[1:2,1:1] on [1,2] of (−1,1);
integer i, t1,t2;
par if (abs(sum(overweigh(M1,A))) + abs(sum(overweigh(M2,A))) +
  abs(sum(overweigh(M3,A))) + abs(sum(overweigh(M4,A))) < 4*t1
  then A:=0
parend
comment 1: threshold is performed on the gradient; t1 is the given threshold
  endcom.
for i:=0 step 1 until 63 do
  begin BIN := 0; par if A = i then BIN := 1 parend;
    HIST [i] := sum (BIN)
  end;
comment 2: to each element of the array A having a given grey value, a
  1-element is correspondingly assigned in the array BIN. Finally
  the sum of these elements in BIN is computed and stored in the
  ith element of HIST. This is done for all the 64 grey values
  endcom.
t2 := 0;
while HIST[t2] ≤ HIST[t2+1] do t2 := t2+1;
while HIST[t2] ≥ HIST3(t2+1] do t2 := t2+1;
comment 3: In order to detect the valley of the bi-modal histogram the central
  minimum must be located. This is done finding the second sign
  inversion when comparing the contents of adjacent t2-locations in
  HIST. The index t2 found in this manner corresponds to the
  threshold of the image endcom.
par if A ≥ t2 then A := 1 else A := 0 parend;
end;

ACKNOWLEDGEMENTS

We thank G. Alfieri, P. Amatruda, N. Ansanelli, M. D'Ambrosio, P. Mastroserio and S. Napoli who are collaborating in writing the PIXAL Compiler.

REFERENCES

1] Arcelli, C., Cordella, L. and Levialdi, S. (1975). "Parallel thinning of binary pictures". Elect. Lett. 11, 148–149.
2] Wood, A. (1977). CAP4 Programmer's Manual. Image Processing Group, University College London.

[3] Duff, M. J. B., Watson, D. M., Fountain, T. J. and Shaw, G. K. (1973). "A cellular logic array for image processing". *Patt. Recog.* 5, 229–247.

[4] HP Algol Programmer's Reference Manual, HP C2116–9072 (1974). Cupertino, California 95014.

[5] Kirsh, R. (1981). "Languages for manipulation of image data structures". Proceedings of the Japan–USA Seminar, Research Towards Real Time Parallel Image Analysis and Recognition, Tokyo, 1978.

[6] Kruse, B. (1976). "The PICAP picture processing laboratory". Int. Rep. No. LiTHISY-I-0096, Dept. of Elect. Eng., Linkoping Univ., Sweden.

[7] Kulpa, Z. and Novicki, H. T. (1976). "Simple interactive processing system 'PICASSO–SHOW'". Proc. 3rd IJCPR, Coronado, USA, pp. 218–222.

[8] Levialdi, S., Maggiolo-Schettini, A., Napoli, M. and Uccella, G. (1978). "Considerations on a language for picture processing". Proc. IVth Polish–Italian Symp. on Bioeng.: Patt. Recog. on Biomedical Objects, Ischia, Italy.

[9] Levialdi, S., Maggiolo-Schettini, A., Napoli, M. and Uccella, G. (1981). "PIXAL: a high level language for image processing". In *Real-time parallel computing*, Onoe, M., Preston, K. Jr. and Rosenfeld, A. (eds). Plenum Publishing Co., New York, pp. 131–143.

[10] Nagel, R. N. and Rosenfeld, A. (1973). "Steps towards handwritten signature verification". Proc. 1st IJCPR, Washington D.C., USA, pp. 59–66.

[11] Stein, J. H. (1963). "Program description of PAX, an IBM 7090 program to simulate the Pattern Articulation Unit of ILLIAC III". Int. Rep. 151, Digital Comp. Lab., Univ. of Illinois.

Chapter Seven

A High-level Language for Constructing Image Processing Commands

D. M. Balston

1. INTRODUCTION

As the technology associated with image processing progresses, so the systems developed by engineers and computer scientists leave the laboratories and begin to be used by experts from other disciplines. The first generation of these systems was little more than the engineers' interpretation of what the experts required. Processing was not very flexible and feedback by way of user satisfaction or dissatisfaction could only be utilized by designing and building another system. Now, however, the second generation is beginning to appear. These are more general purpose systems and allow the users to become more fully involved with the processing they are undertaking. Instead of the engineers becoming instant experts in, say, cytology, the cytologists are becoming better versed in image processing principles. This is as it should be, but maximum advantage will only be gained if the experts can learn to make best use of the systems in the limited time available to them. They do not for example want to know about TV frames, refresh memory, registers and addresses. They do, however, want to make the system perform the function they have just thought of rather than the one they told the designer they wanted. This implies a sophisticated and complex command structure or a facility for generating new commands fairly easily. This chapter describes the latter approach as applied to a system designed to aid in the interpretation of earth resource images.

An ever increasing amount of pictorial data is being collected about our environment and highly trained scientists are needed to interpret the images and unlock the information they contain. Any assistance which can be given to these specialists to speed up their methods and extend their techniques has an important part to play in both the conservation and exploitation of the earth's natural resources. The IDP 3000 is an interactive image processing system designed primarily to help Earth Resource scientists learn more about the capabilities of electronic image processing [1].

The design of the IDP 3000 encompasses the following points:

(i) The system is modular to cater for the varying requirements of different customers;

(ii) The processing is flexible to allow users to experiment and develop new techniques;

(iii) Additional facilities can be added without severe additional cost;

(iv) System response times are faster than human reaction times for processes which require interactive optimization;

(v) The facilities can model existing methods of interpretation;

(vi) The processing is digital to ensure consistency and repeatability.

Of these features, item (ii) is particularly pertinent to this chapter. Normally a system is designed to meet a set of requirements agreed between the designer and the user. In this case no such requirements existed or could be devised. The earth scientists had no experience of using electronic aids of the type proposed, and could only describe their existing methods of analysis and define the areas which they would like automated. Thus the system had to be flexible. It had to provide general purpose arithmetic and logical facilities, and it had to be capable of modelling the existing methods of analysis in order to provide a "point of contact" between the users and the system. Its major role was one of experimentation however, and to fulfil this, it had to be capable of undertaking a range of initially undefined processes.

This flexibility is however, a two-edged weapon. The more flexible a system is, the more difficult it is to learn to use it, and in this case in particular, it was essential that the users be introduced gently to the concepts and facilities of digital image processing. The problem was solved by introducing a small minicomputer between the user and the processing. This interaction control processor interprets the user's commands and sets the relevant hardware functions. It can be programmed to provide interpretation facilities tailored to a specific user and an initial basic set of commands has been provided to cover the common facilities requested by the earth sciences community. It was essential however that those users who became deeply involved with the system should be able to generate their own commands and not rely on the designers of the system to program the control processor. For this reason a high-level language, the Instruction Forming Language (IFL), was devised. This is not a command language, but a language for producing the commands available to the operator. Equally it cannot be considered an image processing language; the facilities provided are those which relate to the hardware rather than the task the hardware is performing. It is felt however that its inclusion here might help to bridge the gap between the existing command languages and the machine independent languages which are so much sought after.

2. SYSTEM STRUCTURE

2.1 General description

Figure 1 shows the main system elements of the IDP 3000. Digital electronic technology is used throughout, the only analogue components being the television camera and monitor. It should be particularly noted that the system is modular and a wide range of configurations can be achieved.

Images are input either via a standard monochrome television camera or from a magnetic-tape unit, the latter being attached to the host computer (typically a large minicomputer) or to the interaction control processor (ICP). The ICP assists the operator to select his processing, provides algorithms for his guidance, interprets his commands and controls the different types of processing provided. Processed images are displayed to the user on a high-resolution colour television monitor.

2.2 The image store

The central element of the system is a specially designed solid-state digital image store capable of providing standard 625 line television signals to the interactive processor and slower data rates to the other processors.

The store is modular in construction comprising up to 16 channels each providing an image of 512 × 512 pixels. Each image point can contain up to eight intensity bits. A single board provides a 1-bit 512 × 512 image and thus the intensity quantisation associated with each channel can easily be modified by adding or removing these boards. The system currently in use at the Royal Aircraft Establishment, Farnborough, comprises five channels each of eight bits. This represents 10M bits of memory operating at up to 10 MHz.

As well as a main store controller which provides centralized timing facilities and television synchronizing signals, each channel has an autonomous control capability. This allows a channel to shift its image horizontally and vertically relative to the others and to operate in a different mode from them. This means that while several stores can be supplying data to the interactive processor and display unit, others can be receiving data from the host computer. In this way, even if time-consuming tasks are being undertaken (in software, for example),

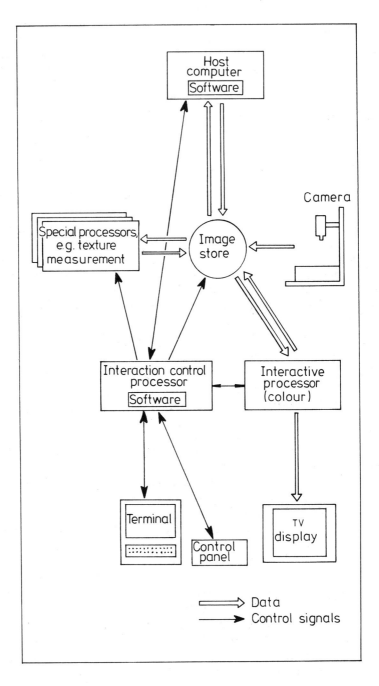

Fig. 1. The IDP3000—basic structure.

the interactive functions are still available and thus maximum use can be made of the system.

2.3 The interactive processor

The processor is modular and comprises a number of distinct functional elements which operate in a pipeline mode and can be combined in almost any order. A novel and very powerful feature is provided by associating control signals with the image data as they pass along the pipeline. In this way, individual control of each pixel can be achieved, thus allowing different parts of the image to be processed in different ways. If desired, one or more bits of a stored image can be used as control signals, and this allows very complex switching patterns. Many of the processing elements are available in quad-rupilcate, thus facilitating comparison of different processes during an optimization task.

A typical configuration of the interactive processor is shown in Fig. 2. Four parallel channels are provided to process the images, while a number of other functions provide quantifying information to the ICP. The names of these functions are generally self-explanatory, but the four element-types comprising the in-line processing require amplification. Image ratioing is provided by the ratio unit which has two inputs, A and B, derived from any store channel. The unit provides either A, A $\div$ B or A $\times$ B. The vector unit provides a weighted sum of four inputs, while the contrast function generator provides a 256 entry table which maps the imput to the output. The tinter/cursor allows a colour to be added to, or to replace, the image data.

2.4 The interaction control processor (ICP)

The ICP comprises a small minicomputer, floppy disk and direct memory access (DMA) channels.

The ICP undertakes the tasks of:
  (i) interpreting and executing operator commands;
  (ii) calculating and specifying real-time control signals.
In order to match human response times, the ICP must be able to interpret and execute commands in 0.1 s. This is a fairly stringent requirement given the complexity of some commands.

Much of the interaction is undertaken by means of a special control panel. This comprises a co-ordinate entry device (typically a joystick), a numeric display and a number of switches. These switches are two-way, centre-return and allow an incremental control method to be implemented. The function of

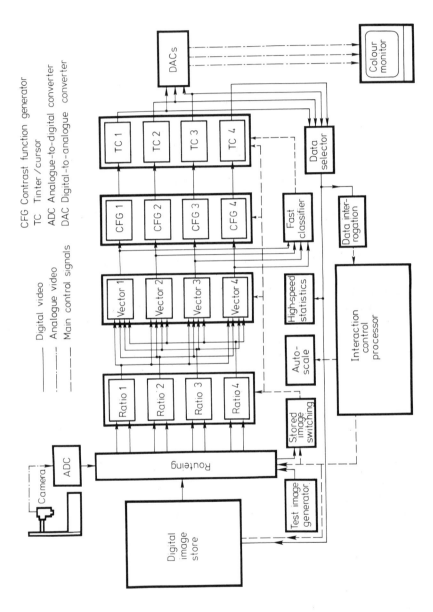

CFG   Contrast function generator
TC   Tinter / cursor
ADC   Analogue-to-digital converter
DAC   Digital-to-analogue converter

——— Digital video
—·—·— Analogue video
— — — Main control signals

Fig. 2. A typical configuration of the interactive processor.

each switch at any instant is determined by the current command which is requested on a standard visual display unit (VDU). The joystick controls a cursor which can be used to define areas of the image for special analysis or to generate outlines of interest to the user. The numeric readout can display, for example, cursor co-ordinates and image intensity values. The ICP can also generate graphical information for display on the colour monitor. This might be, for example, the current function in a contrast function generator or a histogram of image intensity values.

The final use of the ICP is for the storage of successful processing parameters for subsequent recall at a later date. Using this facility, a bank of standard processes could be derived and maintained.

3. THE ICP SOFTWARE

The software in the ICP comprises an operating system, a command interpreter and applications programs (commands).

3.1 The operating system

The operating system is a large, memory resident program that handles transfers to and from the VRP, services a computer–computer link, maintains and updates the monitor display, decodes the control panel, services the keyboard and printer interrupts, decodes command strings, handles the disk drives, and schedules programs. It responds to the hardware's real-time requirements generally making them transparent to the user program, and defines the format of the image to be displayed on the TV screen. It must process operator requests (from both the keyboard and control panel) preferably within 0.1 s to give a truly "interactive" mechanism, although this is subject to the complexity of user programs and the number of them active at any one time. It provides feedback to the user via the system VDU, the control panel and the image display.

The operating system supports a facility for simulating a string of keyboard commands in the "MACRO" subsystem. These files may be generated at run time or extracted from floppy disk. Communications and data links with the host computer are supported allowing the user to extract images (or parts of images) from the host's database and store them in the IDP image refresh memory; similarly images held by the hardware may be sent to the host.

Keyboard commands must be decoded by the software, checked for syntax errors and then performed; two command types are used:

(i) direct—for which an internal routine will execute immediately, e.g. create a macro, list programs in memory.

(ii) indirect—in which a user application will be flagged for future inter-
pretation/execution (if the program is not already memory resident
then it will be brought down from the floppy disk).
Tables 1 and 2 give the range of direct and indirect commands currently
available.

Table 1
Direct keyboard commands

Command name	Function
C	Change common word
CLRPR	Delete all programs from memory
CN	Change current and access next common word
DELAY	Delay execution of keyboard or macro command by given number of TV frames
DELETE	Delete program/macro from memory
EDIT	Create macro
ERRORS	Print DMA/DMT error count
FILE	File a macro on disc
INIT	Initialize system
LISTC	List current programs
LISTF	List programs in memory
LOAD	Load program/macro from paper tape reader
N	Access next common word
P100^a	Set communications mode, ICP
P1P300	Set communications mode, ICP-Host
P300	Set communications mode, TTY-Host
R	Access common word
SHUTDN	Shut down disk operating system
SLIST	List macro
STARTUP	Start up floppy disk
STATUS	Report system status
TRANSF	Initiate Prime 300 image transfer routines
USE	Select tracker ball/joystick for cursor control

^{a} The ICP is a Prime 100 and the host a Prime 300 in the RAE System.

3.2 The command interpreter

The command interpreter is also memory resident and is called periodically by
the operating system. Its purpose is to interpret and execute those applications
programs which have been flagged as active by the operating system. Two keys

Table 2

Indirect commands

Command name	Function
ASSIGN	Specify split screen arrangement
CCOL	Adjust colours in pseudo colour display using switches
CHTINT	Change colours in pseudo colour display
CLASSIFY	Multispectral classification
CONTOUR	Density contour
CURSOR	Generate, adjust and control cursor
ERASE	Erase refresh memory channel
EXPAND	Enlarge cursor
HIST	Measure, display and manipulate intensity histogram
HOLD	Store system parameters on floppy disk
INPUT	Select source images from refresh memory, Camera or Test Image Generator
LABEL	Add labels to display
LOCUS	Generate locus of cursor movement
MIRROR	Display image mirrored in X
MOVE	Move cursor or split screen facility
MULT	Multiply two images
OFFSET	Scale an image
OUTLINE	Outline split screen region
OPTI	Optimize colour space rotation
PATTERN	Generate test pattern
PROCESS	Specify split screen for processing
RATIO	Divide two images
REFRESH	Set refresh memory channel to refresh mode
REMOVE	Modify split screen arrangement
REPLIC	Replicate part of displayed image
RESTORE	Restore system parameters from floppy disk
ROTATE	Rotate colour space
STOP	Inhibit cursor movement
STI	Select test pattern
STORE	Store displayed image in specified refresh memory
SCALE	Measure screen area in scaled units
SLICE	Density slice (pseudo colour display)
STRETCH	Contrast enhance
TRANSFER	Transfer image between IDP 3000 and Prime 300
VECTOR	Linear combination of up to four images

are associated with each applications program. A priority key, which defines the order in which programs shall be executed, and a pass key which defines the number of times the program shall be executed. A pass key of zero implies repeated execution until another program (or keyboard command) causes it to stop.

3.3 The applications programs

The applications programs are held on floppy disk and loaded into memory when first called. They then remain in memory until deleted by the user. This facility allows frequently used commands to execute with little or no delay, because they will usually remain in memory during an interpretation session. The programs are written in the high-level language IFL and "compiled" in the host computer to produce an object module. This contains cross references to the operating system and specialized code which is interpreted by the interpreter.

4. THE INSTRUCTION FORMING LANGUAGE (IFL)

4.1 Requirements

It will be seen from the preceding sections that the IDP 3000 hardware is complex, it has a real time element (the TV frame), it is used by operators who range in experience from the technologically naive to the sophisticated, and it has a range of roles to fulfil.

The concept of the Instruction Forming Language has been incorporated in the design to satisfy the following requirements:

(i) Separate the user from the real time aspects of the hardware.

(ii) Separate the user from the hardware register addresses etc.

(iii) Facilitate the introduction of changes to the basic command set, thus allowing the man–machine interface to be rapidly tailored to the users.

(iv) Allow the more experienced user to construct his own commands.

(v) Provide the user with a frame count facility, thus giving him control of the display period for specific screen configurations.

(vi) Support different configurations of hardware.

(vii) Supply standard logical and arithmetic functions including fixed and floating point representation.

(viii) Include a trace facility for debugging.

4.2 Description of IFL

IFL provides the means by which the basic interaction between the user and the IDP operating system is defined. The operating system contains all the functions which interact with the hardware and these can be called from IFL as external subroutines. Since Basic is the most universally accepted language among the scientists for which the IDP 3000 has been designed, IFL resembles Basic in most respects. Thus it is anticipated that the level of programming experience necessary to understand and modify the commands should be minimal.

Speed of response and ease of debugging are other important features and these are achieved by "compiling" the IFL source into an intermediate object code which is then executed in an interpretive fashion. This is an identical approach to that employed by Pascal for the same reasons.

The main features of IFL are summarized below:

(1) Easy to learn because of its similarity to BASIC.

(2) Interpretive approach gives run-time diagnostics.

(3) Number formats—16 bit integer
        —16 bit pseudo fixed point
        —16 bit pseudo fixed point (9 bit integer,
                              7 bit decimal)
supported directly, plus a floating point package to handle large numbers.

(4) Output statements that format the output and print on the VDU or transmit to the host computer.

(5) Efficient array handling.

(6) Automatic generation of a pseudo operation code to minimize execution time.

(7) Ability to start and end other programs, even if not in memory.

(8) Automatic transfer of program parameters.

(9) Large number of routines available to the programmer.

An IFL program has the same form as a Basic program, namely a declaration section followed by the program body. In addition however, an identification section is needed in IFL. This precedes the declarations and names the program (this will be the command name which is recognized by the operating system) as well as providing pass key and priority key information.

The declaration section must identify the variables used. These may be local to the program or common to the system and can be integer or pseudo fixed point, single variables or arrays. Any variables not declared are assumed to be in the system common area. A data statement allows the definition of the initial values of variables.

The program body comprises a sequence of labelled statements. The labels are integers in the range 1 to 99999 and statements are executed in label order.

The following facilities are identical to those provided by Basic:

LET, GOTO, IF, GOSUB, RETURN, INPUT, PRINT, REM,

The following are facilities which are modifications of Basic:

(*a*) *Variable declaration:* The DIM statement is extended to cover the declaration of variables and to implicitly specify the type.
Thus,
DIM—declaration of integer variable or array;
FDIM—declaration of pseudo fixed point variable or array;
COM—declaration of common integer variable or array;
FCOM—declaration of common pseudo fixed point variable or array.

(*b*) *Data initialization:* The DATA statement takes the form familiar to Fortran programmers.

DATA<variable>/<data list>/

(*c*) *Program termination:* The END statement can be used to terminate the execution of any other currently active program

| END | Terminate this program |
| END<program name> | Terminate another program |

The following are additional facilities not associated with Basic.

(*d*) *Starting another program:* In just the same way that one program can terminate another it is possible to start a second program. When the interpreter meets the START<program name> command it simply inserts <program name> in the current program list.

(*e*) *Calling system subroutines:* This facility provides the most convenient method by which the IFL programmer can interact with the hardware. Every hardware element has one or more subroutines associated with it. These subroutines are written in Fortran and Assembly code and reside in the operating system. The subroutines themselves do not directly interface with the hardware, this is achieved by the operating system via a map of all hardware registers which resides in Common. Thus the subroutines set up the relevant common variables. The programmer can of course directly access these variables but he does need to know the details of the hardware rather more intimately than if he uses the subroutines. There are currently 39 subroutines which control the hardware and a further 29 which control I/O to the VDU, keyboard and host computer and provide the floating point package and other general purpose utilities.

The syntax of the CALL statement is

CALL<subroutine name><subroutine parameters>

(*f*) *Comparing two arrays:* The CMP statement provides a simple way of comparing sections of two arrays. For example, the statement

CMP(<array 1>,<array 2>,<expression>)<label 1>,<label 2>

will evaluate the <expression> and compare the two arrays over <expression> number of words. If the two arrays are equal control will be transferred to <label 1>, otherwise <label 2>.

(*g*) *Logical functions:* The statement forms OR, XOR, AND and SHIFT operate on the bit patterns of words.

OR<variable 1> = <expression 1>,<expression 2>
XOR<variable 1> = <expression 1>,<expression 2>
AND<variable 1> = <expression 1>,<expression 2>
SHIFT<variable 1> = <expression 1>,<expression 2>

In the first three instances the expressions are evaluated, their bit patterns combined using the chosen function and the result placed in <variable 1>. The SHIFT statement right shifts the result of <expression 1> a number of times defined by <expression 2>. If <expression 2> is negative the shift is leftwards.

(*h*) *Restarting a program:* In the same way that a program can be started or terminated from another program, it is also possible temporarily to suspend the execution of a program and restart it from the same or another point in the program flow. The SUSPEND statement simply removes the program from the current program list. Execution of a RESTART statement prior to the suspend will however set the entry point to the program so that when it is next called (either from the keyboard or from another program) it will start at the newly defined position. The execution of an END statement is required to return the start address to its original value. Thus RESTART <label> will cause the next call to the program to start from <label>.

(*i*) *Tracing errors:* A TRACE<expression> statement can be inserted at any point in the program. On interpretation the <expression> is evaluated: if it is non-zero, trace output is produced. If the <expression> is positive the trace output is printed one line at a time, the system waiting for the user to input a carriage return before executing the next instruction in the program. If the <expression> is negative, the output is produced a page at a time.

The trace output gives the program counter value, the pseudo-operation code and the accumulator value in integer and fixed point, all relative to the start point of the program being executed.

5. A PROGRAMMING EXAMPLE

5.1 Introduction

The program chosen as an example of IFL is one which provides the user with real time control of a piece-wise linear contrast enhancement facility. Contrast enhancement is achieved by loading a suitable intensity mapping function into one or more of the CFG elements shown in Fig. 2. In this example the mapping function is to be derived from the settings of up to eight single axis centre-return switches. These switches are software assignable and can be interpreted by the programs in a variety of ways. They are polled by the operating system at regular intervals and the current value of each is increased, decreased or left unchanged depending upon the position found.

5.2 Program specification

The program/command shall be called STRETCH and shall have the following forms,
   (i) STRETCH<channel name>CLEAR
   (ii) STRETCH<channel name>
   (iii) STRETCH UNITY
   (iv) STRETCH OFF.
Form (i) shall set the switch variables to initial values which give a unity transfer function. This clears any previous settings. The <channel name> can be RED, GREEN, BLUE, MONO, or ALL, the term MONO referring to the fourth channel in the system and the colours to the channels which drive the equivalent display primaries. The STRETCH command shall read the switch values on every pass. Each switch shall be assumed to identify the output intensity at 32 input intensity steps from the previous switch. The command shall calculate a piece-wise linear function between all "active" switches. A switch is deemed "active" if its value exceeds that of the previous "active" switch (this limitation ensures a monotonic contrast function).

   The second form of the command is identical to the first except that the switch settings are not initialized before use. The third form sets the contrast function to unity and terminates the command. The fourth form turns the command off (i.e. disables the switches) but leaves the functions set in hardware. All parameters can be abbreviated to their first two letters.

5.3 The program

The IFL program which satisfies this specification is shown below.

```
10                      PROGRAM STRETCH, 0, 1
20      REM
30      REM             CONTRAST STRETCH
40      REM             PARAMETERS ARE:
50      REM             STRETCH RED
55      REM                     BLUE
60      REM                     GREEN
70      REM                     MONO
75      REM                     ALL
80      REM                     UNITY
90      REM
92                      DIM CM(7),A(8),B(8),I,J
94                      DATA CM/'ALREGRBLMOOFUN'/
96                      DATA B/0,0,0,0,0,0,0,0/
98      REM
100                     IF (PM(4)−'CL') 170,110,170
110                     LET I=1
120                     LET J=31
130                     CALL STSW(I,J,0,255,0,0)
140                     LET I=I+1
150                     LET J=J+32
160                     IF (I−8) 130,130,170
165     REM
170     REM     DECODE PRIMARY COMMAND
175     REM
180                     LET I=0
190                     IF (PM(1)−CM(I+1)) 200,240,200
200                     LET I=I+1
210                     IF (I−7)190,220,190
220                     PRINT 'ILLEGAL COMMAND'
230                     END
240                     IF (I−5) 270,230,250
250                     CALL RSU(−1,0,FS)
260                     END
```

270	LET J=0
275	LET J=J+1
280	CALL RDSW (J,A(J))
290	IF (J−8) 275,300,275
300	CMP (A,B,8) 330,310
310	CALL MOVE (A, B, 8)
320	CALL RSU(1,I,FS)
330	RESTART 270
340	SUSPEND

The first statement gives the program name followed by the pass key and the priority key. In this instance the pass key of zero tells the system that the program must be retained in the current program list after is has been executed. Statements 20 through 90 are comments outlining the program.

Statement 92 declares three local arrays CM, A and B and two local variables I and J. In the following two statements, CM is set, two ASCII characters per word, to be the characters necessary to decode the command parameters and B is set to zero.

Statement 100 introduces the common array, PM, which contains the command parameters. Command parameters are limited to two six character ASCII terms and sixteen numeric terms in that order. The operating system on decoding the command from the keyboard, places the first ASCII term, if present, in PM(1) through PM(3), the second in PM(4) through PM(6), and the numeric terms in PM(7), PM(8) etc. Statement 100 tests the presence of the CL parameter in the parameter list. If it is not present, control jumps to statement 170. If it is present the switches have to be set to values at intervals of 32 along the line 0 through 255. This is achieved by statements 110 through 160. The system function STSW sets switch I to the value J. The other arguments define the characteristics of the switch variable. The 0, 255 define the upper and lower limits on its value. The fifth argument is a flag which specifies that the value should hard limit at the end of the range rather than "rotate" back to the other end of the range. The final argument controls the sensitivity of the function relating the switch position to the increase or decrease in its value. An argument of zero implies no change to this function from that previously used.

Statement 160 counts the number of switches that have been set and jumps on to statement 170 when all eight have been set up.

Statements 180, 190 and 200 convert the initial ASCII parameter in PM(1) to an equivalent numeric value which is set in the variable I. Thus AL maps to I = 0, RE to I = 1 and so on. If none of them are present statement 210 detects this, passes control to statement 220, an error message, and the program terminates at statement 230.

If the OF parameter is detected (I = 5) the program terminates at statement 260 after calling the system function RSU. This is the function which effectively sends data to the hardware registers of the contrast function generator (CFG). The arguments associated with this particular call, result in a unity function being sent to the CFG's in all channels. The argument FS defines a further refinement of the hardware which cannot be amplified here due to lack of space.

Statements 270 to 340 perform the function once the command has been established. Statements 270 through 290 call the system function RDSW for each switch resulting in the current value being assigned to A(J) for switch J.

Statement 300 compares A with the last set of values, B, moves the values from A into B if they are different, and calls RSU with the first argument set to indicate that the function should interpolate between the switch values maintaining the monotonicity criterion. The second argument, I, defines the channel to which the data is sent. Finally the program start point is set to statement 270 and the program suspended.

It will be seen from the foregoing description that the majority of the program is concerned with command parameter interpretation. That is commonly the case because users have very strong ideas about the format of commands, about what prompts they want, what abbreviations are acceptable and so on. In the example the subroutine RSU undertakes the bulk of the actual work associated with the command. RSU is capable of:

  (i) Setting a unity transfer function;

  (ii) Setting a logarithmic transfer function;

  (iii) Setting a zero transfer function;

  (iv) Interpolating monotonically through up to eight points;

  (v) Accepting a 256 entry table.

Supportive system functions are routines for interpolating between two points, interpolating through one point given a slope, floating point routines and trignometric routines. Thus the amount of programming required for a new command is generally fairly small.

6. CONCLUSIONS

The system has been fully operational at R.A.E., Farnborough for two years and is proving very successful. Surprisingly, in view of the original requirements, very few changes have been made to the command set by the users. Perhaps the time-scales have been too short or the range of commands has been wide enough to satisfy the variety of users which has had access to the system. It is therefore impossible to claim that the language has achieved its objective.

On the other hand a substantial extension to the system software has been written by members of the design team. A range of spatial processing commands which allow neighbourhood processing and convolution filtering to be undertaken has been written entirely in IFL. Only one of the team writing this software had had experience of IFL when writing the initial command set. Thus it can be claimed that the language proved easy to learn and sufficiently flexible to allow the generation of some very complex routines involving sequences of operations occupying successive TV frames.

REFERENCE

[1] Balston, D. M., and Custance, N. D. E. (1979). Inst. Phys. Conf. Ser. No. 44: Chapter 8, pp. 287–302.

Chapter Eight

A Generalized Support Package for Image Acquisition

W. Black, J. F. Harris and T. Clement

1. INTRODUCTION

For the past seven years, two precision CRT film scanners (PEPR) have been used at Oxford for a variety of image processing applications [1]. Both systems were originally developed for bubble chamber film processing [2] and hence several of the applications have naturally been in the area of line structured data. Much work has been done in the area of digitizing line trace records for rain [3] and river level recorders and several other similar applications [4]. Also, research work is proceeding on the automatic interpretation of thematic maps and engineering drawings [5].

The systems have also been applied to areas other than line structured image processing; the main efforts here have been in processing film images of a radar PPI screen showing shipping in the English Channel [6], and some work on the automatic planimetry of irregular areas on land usage maps.

Software development for the bubble chamber film scanning system dates from 1966 and is FORTRAN based. The later diversification into other application areas has resulted in a rationalization of software for image scanning and interpretation. The aim of this chapter is to describe how we have used FORTRAN to provide a generalized support package for image acquisition.

The package is written to run on a DEC System 10 computer for both Oxford PEPR machines and has been in use for the last four years. A version is being undertaken for a LASERSCAN FASTRAK [7] device on a DEC VAX 11/780. The type of scanning device principally addressed by the package is of high resolution and can randomly access the picture as the data store, e.g. the typical resolution of 10 μm over a 50 mm $\times$ 50 mm frame of film is equivalent to 25 million pixels. Although the package would be suitable for use on lower resolution devices, we have in no way attempted to support parallel processing hardware.

2. TOWARDS STRUCTURED SCANNING SOFTWARE

While developing software for many varied applications, several objectives become apparent: first, the need for device independence so that programs written for or developed on one scanning machine may be transferred easily onto another; secondly, the avoidance of duplicated effort, especially in areas such as calibration or of driving the hardware, leaving the applications programmer free to concentrate on the higher level problems; thirdly, the need to be able to develop new applications quickly to improve the cost effectiveness of short production runs. To meet these objectives a new software package was written giving a structure as shown in Fig. 1.

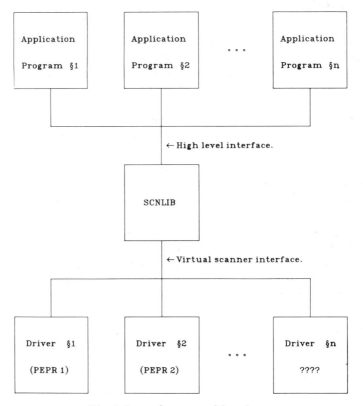

Fig. 1. Layered structure of the software.

The upper interface is well defined [8] and permits the applications programmer to address his chart or picture in units of his choosing which are suitable for the problem being solved. It also provides him with a variety of comprehensive scanning sequences over areas or along lines of arbitrary size. The use of this interface permits the speedy development of new programs.

The scan library (SCNLIB) is written in FORTRAN and carries out all mappings and calibrations, performs scans over areas, along lines, and at points and returns data in a form convenient to the user, thus meeting the second objective.

SCNLIB "drives" a hypothetical scanning device or "virtual scanner". The definition of this scanner is again precisely specified [8]. It is designed from an abstracted set of properties of real scanners to facilitate its mapping onto them. Its underlying co-ordinate system may require calibration and scans along lines may be of restricted lengths. However its operation is asynchronous, permitting SCNLIB effectively to overlap its computations with scanning. The virtual scanner gives device independency satisfying objective number one.

For each device available a driver must be written, probably in a machine dependent assembly language. It must simulate the virtual scanner on the particular device. Although a fairly small module, specialist knowledge is required to write it. However, once done for a device, it need never be redone for new applications.

3. THE VIRTUAL SCANNER INTERFACE

The virtual scanner is considered to be an idealized data acquisition device. A device driver will normally be required to translate "commands" to the virtual scanner into (possibly sequences of) commands for the real piece of hardware.

The scanner has access to a two-dimensional image and imposes on this an orthogonal co-ordinate system: "(u,v) space". This access is by means of a "probe" (CRT spot, laser beam or photo cell). The probe has a finite size and this imposes a maximum resolution on the scanning process. The probe may sample the image and return an "intensity" digitized into a suitable range, or may compare the image with a threshold which has been previously defined.

Commands to the virtual scanner fall into six groups: (1) initialization and termination; (2) set parameters; (3) ask parameters; (4) ask environment; (5) start scans; and (6) retrieve results.

Group 1 is self explanatory. Parameters which may be set by commands in group 2 include: the (u,v) co-ordinates for the initial probe position; scan direction; threshold for probe comparisons; and scan length.

All parameters which may be set by group 2 commands may also be read by corresponding commands in group 3. Note, however, that due to the discretization imposed on the (u,v) space by the finite probe size, digitization of intensity etc., the values returned may be rounded from those actually given in the set command.

Group 4 commands return data on the nature of the real underlying device, e.g. maximum linear scan length, maximum resolution, number of discrete addressable points, intensity range. These allow for a tailoring at run time by a higher-level program to a variety of different scanners.

Scanning may be at a point or along a line of a specified length parallel to one of the axes of (u,v) space. Combined with the two methods of probe data acquisition described above, this gives four possible "actions" for the virtual scanner, viz. intensity at a point, intensities along a line, comparison with threshold at a point, and comparison with threshold along a line. These make up the group 5 commands. The execution of the actions is considered to be asynchronous, i.e. when an action is started the data from the previous action may be analysed and parameters set for the next action. Synchronization and retrieval of results are provided for by the group 6 commands.

We have implemented the virtual scanner as two FORTRAN callable libraries of subroutines, one for each of our PEPR's, such that each CALL performs a virtual scanner command.

4. SCAN LIBRARY

Although the virtual scanner mechanism permits device independence, it is not yet a suitable interface for the applications programmer. Many factors suggest this, e.g. inconvenience of the (u,v) space units, distortions of the scanner's view of the image (CRT devices show a strong pin-cushion effect), limited scan length/direction. The scan library is a FORTRAN subroutine library which uses the virtual scanner to access the image and provides an application oriented interface.

First, it supports user chosen spaces. Some examples include millimetres on film, kilometres on a map, and hours against inches of rain on a rainfall chart. The scan library maintains mappings between the chosen space and the underlying (u,v) space. All other calls are in terms of this chosen "chart" space. It is during this mapping stage that calibration procedures are invoked to compensate for any scanner induced distortions, e.g. 5th order polynomial corrections for pin-cushion distortion are used for PEPR.

Secondly, scans may be performed over extended lines, which may be at arbitrary angles, and over rectangles in chart space. The library will interrogate the virtual scanner for the limits of scan lines and hence construct suitable scan patterns which it will interleave with the calibration and mapping procedures. Thus the user, in writing his programs, is freed from the restrictions normally imposed by a real scanner.

Finally, the package contains useful debugging and operator interaction facilities. During the development process, it is helpful to display the results of each step in the total scan strategy. Each data acquisition call in the library has a built in "breakpoint". If enabled, this will display a picture on an interactive graphics terminal associated with the scanner. The graphics package used is GHOST, but the calls are fairly well localized and could be converted to other packages such as GINO. The graphics package allows an operator to indicate a position by means of a tracking cross on the display and this, in conjunction with the debugging system, permits rescans at the same or different place on the image and generally allows a developer to experiment with different scan strategies.

A full user's guide is available giving details of all the subroutines which may be called.

5. SIMPLE EXAMPLE OF SCNLIB USE

Consider that we have located an oval-shaped feature on the image and we wish to obtain a measure of its area. We construct a box about its position and scan across it by a series of lines parallel to the x-axis of chart space. The following code gives an indication of a suitable algorithm (but is not necessarily the best or most efficient):

```
      CALL SETTHR(THR)                    !threshold
      CALL SETASZ(DX,DY)                  !area size
      CALL SETSPC(DYSP)                   !line spacing

      CALL DOACGP
     (XC,YC,0.0,NARR,XARR,YARR,NHT)       !scan over the box

      NHT=MIN0(NARR,NHT)/2                 !number of hits
      AREA=0.0
      IF(NHT.LE.0) GOTO 20                 !if too few

      DO 10 I=1,NHT                        !for all strips
   10 AREA=AREA+DYSP*ABS(XARR(2*I)-
      XARR(2*I-1))                         !add this strip

   20 CONTINUE
```

THR—is the threshold at which to look for hits
(DX,DY)—define the size of the surrounding box

DYSP—is the spacing required between the scan lines
(XC,YC)—is the centre of the box
NARR—is the size of arrays XARR, YARR
NHT—is the number of transitions through the threshold
AREA—is the area required

6. HIGHER LEVEL ROUTINES

The subroutines available to users of SCNLIB perform essentially the functions of image collection, pre-processing, and a limited amount of filtering. There is therefore a further layer above that of SCNLIB whose routines perform the higher level functions of segmentation, feature extraction, recognition etc. The task of making a library of such routines is much greater than that of making SCNLIB since there are, for example, many different algorithms available and some are more applicable to certain situations than others.

It is envisaged that this higher level "chart library" (CHTLIB) will be open-ended, being similar in structure to the NAG library of numerical algorithms. Routines will be classified into subject areas, will be well documented, and will have aids to help the user select the best algorithm for a given task. The general, precisely defined and machine independent nature of SCNLIB lends it to being a suitable and practical tool for the description and interchange of higher level routines and so all contributions will be programmed in terms of SCNLIB calls.

At present CHTLIB is small and contains a fairly general chart edge locating sub-package, a group of small routines for the detection of straight lines from a collection of points over an area, and a general table driven trace follower. Raster to vector and character recognition algorithms are being developed but are not yet incorporated.

7. CONCLUSIONS

Our experience with the software described above shows that FORTRAN is a useful language for image acquisition and processing, provided that it is suitably augmented by libraries of reliable and well designed subroutines. The use of PASCAL or ALGOL 68 would certainly tidy certain constructions syntactically, but we have not yet found problems which are insurmountable in FORTRAN. The key is in the richness, generality and device independence of the subroutines.

REFERENCES

1] Black, W., Clement, T., Davey, P. G., Harris, J. F. and Preston, G. (1979). "The use of an interactive high-precision CRT scanner for image processing". *Inst. Phys. Conf. Ser.* **44**, 308–318.

2] Davey, P. G., Harris, J. F., Hawes, B. M., Story, C. and West, N. (1979). "A four-view automatic measuring system for bubble chamber film". *Inst. Phys. Conf. Ser.* **44**, 67–77.

3] Davey, P. G., Harris, J. F. and Loken, J. G. (1974). "PEPR techniques for reading microfilmed rain charts". Proc. Oxford Conf. on Computer Scanning.

4] Black, W., Clement, T., Harris, J. F., Llewellyn, B. and Preston, G. (1981). "A general purpose follower for line structured data". (To be published in *Patt. Recog.*)

5] Clement, T. (1981). "The extraction of line structured data from engineering drawings". (To be published in *Patt. Recog.*)

6] Preston, G. (1981). "Automatic tracking of filmed radar echoes". (To be published in *Patt. Recog.*)

7] Fulford, M. (1981). "The FASTRAK automatic digitising system". (To be published in *Patt. Recog.*)

8] Clement, T. (1979). "An investigation into techniques for automated data acquisition from graphical images". Thesis, Univ. of Oxford.

Chapter Nine

TAL: An Interpretive Language for the Leyden Television Analysis System*

J. Vrolijk, P. L. Pearson and J. S. Ploem

1. INTRODUCTION

In our laboratory two projects are being carried out based upon the Leyden Television Analysis System (LEYTAS), one concerned with automating the prescreening of cervical smears [1] and the other with the automation of chromosome analysis [5]. We felt that a software system controlling an image processing system intended for use in various fields of biomedical research should fulfil the following requirements. The language should not only be suitable for experienced programmers, but also for cytologists, cytogeneticists, etc., so that they should be able to investigate for themselves, if their questions could be solved by the system. Since human thought processes are not intrinsically equipped to visualize mentally the results of iterative transformations of biological objects, the software system should provide the possibility of displaying intermediate results, execute procedures step-wise, allow an easy adaptation of parameter settings and other programmable aspects, such as type of transformation and manner of operator interaction, whenever necessary. Particularly in the testing phase of image analysis, interaction and flexibility are essential features.

Accordingly, an interpreter language seemed preferable to a compiler-oriented language. Although programs controlled by an interpreter usually run slower than compiled programs, we felt that an interpreter would still permit running programs at the maximum speed permitted by the hardware

*This work is supported by grants of the Stichting Koningin Wilhelmina Fonds and the Praeventiefonds, The Netherlands.

(essentially limited by the video frame rate) in order to avoid unacceptable execution times during experiments involving the analysis of several thousands of microscopic fields.

This chapter first provides an abbreviated description of the hardware configuration of LEYTAS to inform the reader which image procedures are permitted by the hardware. Secondly, the structure of the software system TAL is described in detail and, finally, a programming example is given to illustrate its possibilities.

2. HARDWARE CONFIGURATION

LEYTAS consists of an automated microscope, the Leitz TAS image processor, a grey-value memory for the storage of television scanned images, an LSI microprocessor and a PDP 11/40 minicomputer (both from Digital Equipment). The basic configuration of the system is shown in Fig. 1.

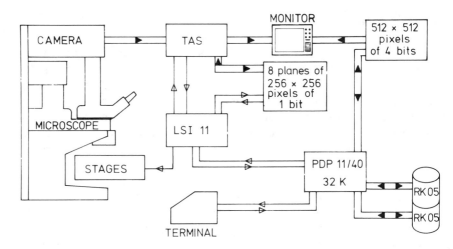

Fig. 1. The configuration of LEYTAS.

The automated microscope is adapted to suit the different applications carried out in our laboratory [4] and is provided with the following functions: switching between absorbtion and epifluorescence optics, the selection of different wavelengths and magnifications using an illuminator and projective rotor respectively, the control of a Zeiss 10 μm stage (300 Hz) for slide positioning and autofocussing in the z-axis. All functions can be controlled directly by a special keyboard or by program commands.

The image to be analysed is fed into the transformation circuits by a

Plumbicon television camera (with a frame rate of 50 Hz). Simultaneously with the scanning, the image can be processed by the TAS and also stored directly in a 4-bit grey-value memory following digitization via an 8-bit AD-converter (10 MHz). The necessary compression from eight to four bits is fully programmable. The grey-value memory consists of four 4-bit memories, each of 256×256 pixels, which can also be used as a single memory of 512×512 pixels by storing four consecutive frames positioned on every second pixel and line relative to each other. Eight-bit grey-value information can also be obtained by storing two consecutive frames and adjusting the AD-converter between the frames. Communication between the PDP 11/40 and the grey-value memory is carried out on the basis of a direct memory access providing the highest transfer rate possible, e.g. approximately 2 s for an entire image.

The TAS itself is a 1-bit picture processor based on a triangular grid. Image transformations are performed by the parallel processing of a small neighbourhood of each pixel sequentially. The basic transformations are erosion (shrinking) and dilation (expansion), which can be defined as:

$$S \ominus B = (x|B(x) \subset S) — \text{erosion of S by B}$$
$$S \oplus B = (x|B(x) \cap S \neq 0) — \text{dilation of S by B}$$
$$\text{where } S = \text{set of pixels,}$$
$$x = \text{element of set S,}$$
$$B(x) = \text{structuring element centred at x.}$$

Also the combinations erosion followed by dilation, called ouverture (opening), and dilation followed by erosion, called fermeture (closing), can be combined in the same scan. The structuring element in the TAS can be varied using the seven points of the elementary hexagon, the centre being one of the points for any suitable subset. The 1-bit source image for these transformations is either a boolean combination of two different thresholded images derived directly from the camera or a boolean combination of two pictures stored in a bitplane or the intersection of both types. The transformed picture can subsequently be stored in a bitplane, whilst features such as area, number of particles, or perimeter can be extracted simultaneously. Thresholding is performed either by selecting an upper and lower bound from the total grey-scale or by a gradient operator in the horizontal direction. Eight bitplanes of 256×256 pixels are available for storing the threshold image or transformed versions of the same. The implementation of bitplanes allows the development of iterative image analysis procedures such as skeletonization. Furthermore, a lightpen unit is incorporated in the system for interactive control and a contour follower module allows the analysis of individual objects. Using the latter module the co-ordinates of the contour points of each object can be stored in the computer, so that it is possible to select objects from an entire field on the basis of image analysis procedures and subsequently to read only those points into the grey-value memory, which are located within the objects. Thereafter the objects can be analysed further by pattern recognition algorithms independently of hardware transformations. A more detailed description of the TAS and its potentialities in image analysis is given by Nawrath and Serra [2, 3].

The entire system is controlled by a PDP 11/40 minicomputer, provided with 28k core memory and 2 RK05 disk drives for data and program storage running under the RT11 operation system. A LSI 11 microprocessor with 24k memory is included in the system to relieve the PDP 11/40 of the stage control, automated microscope and handlers for the hardwired modules of the TAS.

3. THE COMMAND STRUCTURE OF TAL

The software system TAL developed in our laboratory has complete control of LEYTAS and as such, is not only a language, but also a type of operating system. It allows direct control of LEYTAS by means of single commands, the creation, updating and execution of image analysis programs. Other features include facilities for interruption, continuation and single step execution of programs and the provision of limited file-management capabilities for programs written in TAL.

After starting TAL, a basic list of options is displayed on the video terminal. The selection of one of those will result in subsequent lists, until the complete instruction necessary to carry out a single type of hardware use is encoded. Following the direct execution of the required operation the basic list will again be displayed on the video terminal. For example, the selection of the instruction EROSION with size 5 is shown in Fig. 2.

The option list structure was chosen to facilitate control of LEYTAS by non-experienced programmers and others not acquainted with the system. In this way the operator has access to complete control of the system without knowing any syntax rules. TAL checks each instruction directly during encoding, so that it is impossible to generate an illegal one. Since each command is directly carried out and the results of the operation immediately shown on a television monitor, LEYTAS is suitable for developing image analysis procedures interactively.

After starting TAL, the system enters the single command mode. Although most commands can also be applied as program statements, there are a few obvious exceptions, which are only applicable as single commands, namely: EXIT, KEYBOARD LAYOUT, PROGRAM CONTROL and PERIPHERAL CONTROL (Fig. 2). The KEYBOARD LAYOUT option informs the operator which keys of the microscope keyboard are assigned to which microscope and stage functions, whilst the EXIT option terminates the program and returns control to the RT11 operating system.

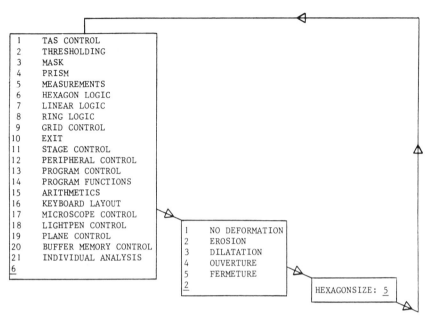

Fig. 2. The selection of the instruction: EROSION: 5.

The PROGRAM CONTROL option contains all the necessary commands to allow the creation, updating and deletion of programs. Following the BEGIN PROGRAM command each instruction is stored in a program buffer, until the END PROGRAM command is given. During the editing a statement number is generated below the basic list of options on the terminal. On the basis of these statement numbers, statements can be deleted or inserted following respectively the command DELETE STATEMENTS or START INSERT STATEMENTS. After the EXECUTE command, the program, which is resident in core, is carried out. During execution, the operator can force the program to continue in single step mode by pressing the (control D) key and can stop execution by pressing the (control E) key. Before continuing execution, by pressing (control E) again, the operator can change any TAS parameter (threshold setting, erosion size, etc.) or bitplane content in the single command mode. We have found that these features are very useful during the testing of algorithms.

Using the PERIPHERAL CONTROL option, it is possible to read, write or list programs written in TAL from or onto various peripherals such as disk, papertape unit, printer and console terminal. (All peripherals supported by RT11 are also supported by TAL.) Two device handlers have been added to

RT11, allowing respectively, the addressing of bitplanes and frames of the grey-value memory in the same way as a printer.

All options from the basic list (Fig. 2) not yet mentioned, can be applied as legal program statements for TAL programs. These instructions can be divided into the following types: microscope and stage functions, instructions to control the buffer memory, TAS functions and program functions.

The microscope functions allow the positioning of shutters, illuminator rotor and projective rotor. For stage control, functions are available to step relatively, to reposition at absolute co-ordinates, to locate the co-ordinates of an operator-defined origin so that the position objects relative to the origin can be defined, and to initialize the stage at its hardware origin.

Instructions for the control of the grey-value memory are: load a frame, display a frame, load a lookup table for the AD-converter, read or write a part of a frame from or into the grey-value memory, and threshold a frame and copy it into a bitplane. Furthermore, routines have been developed allowing the creation of vector plots and histograms. (These routines are also available for the bitplanes.)

The TAS functions control the hardwired modules of the image processor and can be divided into asynchronous and synchronous instructions. The latter are synchronized with the television frame and their execution time takes either one TV cycle of 20 ms or a number of TV cycles for iterative procedures. Examples of these instructions are the storage of an image in a bitplane, the logical intersection between two bitplanes followed by storage (all 16 possible boolean functions between two operands are available) and various measurements. The asynchronous instructions are carried out within a few microseconds and are used to switch on or off the different image transformation units. Examples of those instructions are: erosion, thresholding and ringlogic. The latter allows the selection of various structuring elements for the erosion and dilation transformations. The six combinations, which are necessary to perform a skeletonization, are listed in Fig. 3.

The program instructions are the normal instructions necessary to develop a program, such as arithmetic instructions, input/output (IO) instructions, (un)conditional jump instructions, etc. Unformatted sequential, formatted sequential and unformatted random access IO are supported by TAL. These program instructions can only be implemented during the normal run mode of programs and not in the single command mode. Although these instructions are generated according to the TAL philosophy of instruction selection from lists (an example of the generation of an IF-statement is given in Fig. 4), these instructions can be looked upon as a subset of FOTRAN IV, since they observe the same syntax rules.

As TAL is provided with the option to generate FORTRAN compatible routine calls for the instructions dedicated to LEYTAS, whenever an operator

```
0 0          0 d          d 1
d 1 d        0 1 1        0 1 1
  1 1          d 1          0 d


  1 1          1 d          d 0
d 1 d        1 1 0        1 1 0
0 0          d 0            1 d
```

```
0 - test if corresponding pixel is '0'.
1 - test if corresponding pixel is '1'.
d - don't care condition.
```

Fig. 3. The six structuring elements necessary to perform skeletonization.

```
EXPRESSION: I
CONDITION: GT
EXPRESSION: IFIX(SQRT(3.))
LABEL: 100
```

Fig. 4. The generation of the IF-statement:

IF(I.GT.IFIX(SQRT(3.))) GOTO 100

(The underlined words are generated by TAL.)

lists a program with the FTN or FOR extension, a legal FORTRAN source file is obtained. After compilation, interactively developed programs can thus be run independently from the interpreter and easily implemented in larger pattern recognition or classification programs written in FORTRAN. Normally however, the statements, as dedicated to LEYTAS, are more readily readable than in their FORTRAN form.

The following data types are supported by TAL: LOGICAL, INTEGER and REAL. These data types can be represented by constants, variables and one-dimensional arrays. In addition, bitplanes can be regarded as a data type, although their manipulation is limited to the instructions permitted by the hardware.

4. THE INTERNAL STRUCTURE OF TAL

Each encoded instruction is organized in the memory as follows: the first word contains the length of the instruction, the second a function code and subsequent words the necessary parameters. The highest four bits of each parameter contain a code indicating its type. In the case of a constant, the next word(s) represent(s) its value. In the case of a variable or array element, the other twelve bits contain an offset address to a buffer carrying the names of all variables and arrays. This name buffer is organized as follows. Each variable or array is represented by three words. For each variable the first two words contain the name, whilst for an array the first word contains the number of elements and the next one the name. The third word consists of the identification and a pointer to the relevant part of a buffer containing the values of all the variables and arrays. This double indirect addressing is necessary to enable the disassembly of an instruction, whenever a listing is required. Figure 5 shows the concept of the parameter organization schematically. For the instructions specifically dedicated to LEYTAS only constants, variables, arrays and array elements are allowed as parameters with no expressions. For the FORTRAN compatible statements expressions are allowed. The organization of the encoding of FORTRAN functions is accordingly very similar to the handling of variables. The highest four bits contain again a type identification, whilst the other twelve bits contain an offset to the name buffer, which specifies the name of the function (two words) and an offset (one word) to a jump-table with the addresses of the routines. Figure 6 shows an example of the encoded representation of an expression.

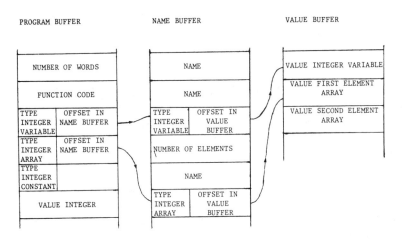

Fig. 5. The internal organization of parameters.

instruction code (octal words)	description
type	

								description
				1	7			– number of words
			4	2	2			– function code
1	0	0	0	2	4		IA	– type: INTEGER array
1	4	0	0	0	2		(	
0	4	0	0	1	6		I	– type: INTEGER variable
1	4	0	0	0	5		)	
1	4	0	0	1	3		=	
0	4	0	0	2	1		J	– type: INTEGER variable
1	4	0	0	4	6		+	– INTEGER addition
1	5	0	0	6	2		IFIX	– REAL to INTEGER conversion
1	4	0	0	0	2		(	
0	6	0	0	2	7		A	– type: REAL variable
1	7	0	0	4	3		+	– REAL addition
0	6	0	0	3	2		B	– type: REAL variable
1	4	0	0	0	5		)	
1	4	0	0	1	6			– indication for end of expression

Fig. 6. The internal representation of: $\mathrm{IA(I)=J+IFIX(A+B)}$.

As only program functions and commands involving the grey-value memory are processed in the PDP 11/40, whilst the others are handled by the LSI, the function codes can be divided into two groups with a corresponding jump-table. On the basis of the function code it is decided whether the variables should be transferred to the LSI or not. The group of LSI functions can be further subdivided into those from which one expects an answer back and those which give no response. Each variable, into which an answer should be stored, is temporarily disabled by changing its type identification in the name buffer. Program execution continues until the blocked variable is encountered again; the program then stops and waits until the answer from the LSI is received and the variable is re-activated again. In this way the interrupt mechanism remains invisible to the user. Figure 7 shows a simplified flow diagram of interpretation of TAL programs.

The LSI encodes the TAS functions and controls the timing of the micro-scope and the stage functions. As there are only a few milliseconds available to encode the functions between two synchronous TAS operations without losing TV cycles (Fig. 8), the encoding of the function is carried out during analysis of the previous instructions by the TAS and are stored in a ringbuffer. As soon as the TAS is ready to accept new commands, the microcodes are called from the ringbuffer and transferred to the TAS. In this way it is possible to transfer

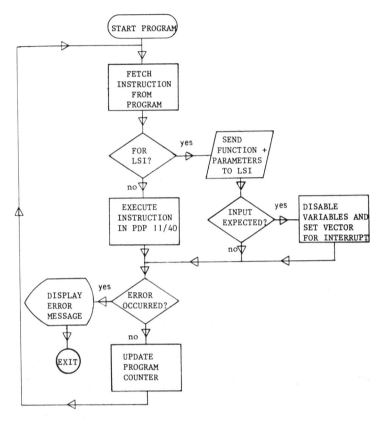

Fig. 7. Flow diagram of the interpretation of programs written in TAL.

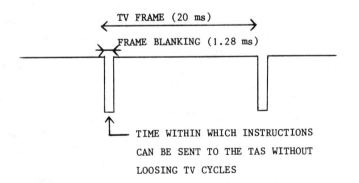

TV FRAME (20 ms)

FRAME BLANKING (1.28 ms)

TIME WITHIN WHICH INSTRUCTIONS
CAN BE SENT TO THE TAS WITHOUT
LOOSING TV CYCLES

Fig. 8. Timing diagram of TV frames.

about 20 microcodes between two measurement instructions and 50 microcodes between two bitplane instructions without losing a TV cycle. For most programs this number of microcodes is sufficient to allow the processing of the TAS at its full speed. Figure 9 shows a flow diagram illustrating the handling of microcodes within the LSI.

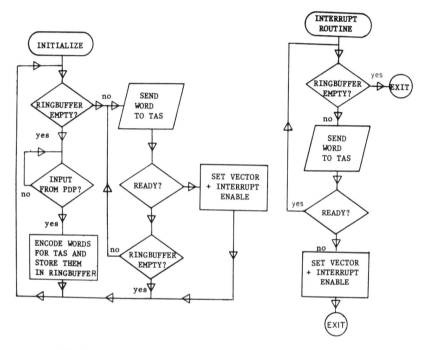

Fig. 9. Flow diagram of the handling of instructions in the LSI 11.

5. A PROGRAM EXAMPLE

The selection of metaphases is probably one of the more important aspects in the field of automated chromosome analysis. According to the definition of a metaphase as a cluster of small objects located relatively close to each other, a metaphase can be selected on the basis of a fermeture. Following the dilation step, the chromosomes cluster together and become an object, usually larger than the nuclei. Following an erosion, the metaphase remains detected, as illustrated in Fig. 10. This initial selection criterion can be used in the first stage of a reliable metaphase selection procedure.

A program written in TAL for the selection of metaphases is partially listed in Fig. 11. In the program each image is stored in the grey-value memory

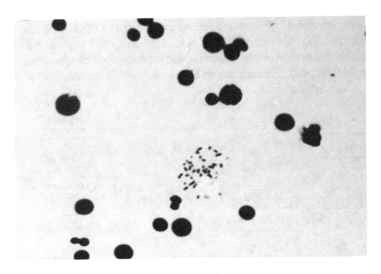

Fig. 10. A metaphase selected using the fermeture criterium.

```
INTEGER*2 X(1000),Y(1000)
K=30
J=1
LOAD LINEAR LOOKUP TABLE: 0,255
DO 20 IY=1,40,1
DO 10 IX=1,100,1
THRESHOLD LEVEL: 50
THRESHOLD RANGE: 35
FERMETURE: 3,7
SYNCHRONIZE TAS & MICROSCOPE
LOAD FRAME: 0
AREA MEASUREMENT: A
X-COORDINATE STAGE: X(J)
Y-COORDINATE STAGE: Y(J)
STEP STAGE X DIRECTION: K
IF(A.EQ.0.) GOTO 10
CALL FIND(X(1),Y(1),J)
10       CONTINUE
STEP STAGE Y DIRECTION: 28
K=-K
20       CONTINUE
STOP
SUBROUTINE FIND(X(1),Y(1),J)

         /
        /
       >
      /
     <
END
```

Fig. 11. The listing of a program for metaphase selection.

simultaneously with testing for the presence of a metaphase. Even if a possible metaphase is found, the stage is moved to the next field. During this stage-movement, the analysis to ensure that the selected object is indeed a metaphase is carried out on the image stored in the grey-value memory. In this way it is possible to overlay stage-movement and image analysis in order to obtain a higher efficiency. A synchronize statement is added to prevent that the next field is analysed before the stage is settled.

6. CONCLUSION

Since TAL has been written in PDP 11 assembler dedicated to LEYTAS, it is not directly transferable to systems based on other computers or image processors. However, as all image analysis functions are FORTRAN callable routines, it is possible to skip the interpretive part of TAL and to develop programs in FORTRAN. Only the addition of a routine to load the LSI part of the TAL system and a handler in the PDP 11/40 to add the correct function code to each routine and take care of the communication with the LSI, are necessary to allow implementation of programs in FORTRAN. Since an LSI is incorporated in each standard TAS, the LSI part of TAL could easily be stored in PROMs. The LSI can then in fact be regarded as an intelligent and integral interface for the TAS resulting in a machine independent, FORTRAN controllable image processor.

REFERENCES

Al, I. and Ploem J. S. (1979). "Detection of suspicious cells and rejection of artifacts in cervical cytology using the Leyden Television Analysis System". *J. Histochem. Cytochem.* **27**, 129–134.

Nawrath, R. and Serra, J. (1979). "Quantitative image analysis: Theory and instrumentation". *Microsc. Acta* **82**, 102–111.

Nawrath, R. and Serra, J. (1979). "Quantitative image analysis: Applications using sequential transformations". *Microsc. Acta* **82**, 113–128.

Ploem, J. S., Verwoerd, N., Bonnet, J. and Koper, G. (1979). "An automated microscope for quantitative cytology combining television image analysis and stage scanning microphotometry". *J. Histochem. Cytochem.* **27**, 136–143.

Vrolijk, J., ten Brinke, H., Ploem, J. S. and Pearson, P. L. (1980). "Video techniques applied to chromosome analysis". *Microsc. Acta*, Supplement 4, 108–115.

Chapter Ten

High-level Image Languages for Automatic Inspection— the Filestructure Problem

L. Norton-Wayne

1. INTRODUCTION

Image processing languages are required to be standardized, transportable, applicable to the widest possible range of task within a given problem area, and at high-level. The level of a language is here defined to be the ratio between the minimum number of machine code instructions required to perform a typical sequence of tasks, divided by the number of instructions required in the high-level language. An alternative and equivalent definition specifies level by the ratio of keystrokes required of the programmer between the two.

Several image processing languages have emerged. These fall into two classes. In the first (of which the languages VICAR [1] and IMPROC [2] are examples), the processing is performed by sequences of subroutines written in standard compiled languages. For VICAR and IMPROC, this language is FORTRAN. The "high-level" language (in VICAR, this is called the command language) comprises a special code which calls (with the minimum number of keystrokes, and without the "high-level" user needing even to know FORTRAN) sequences of FORTRAN subroutines which perform the task desired. These languages are highly transportable, though slight modification is normally needed to accommodate dialect, and to replace the few segments written in assembler to facilitate communication with peripheral devices. It is easy for users to add their own compatible segments to these languages, but they tend to be inefficient in their use of memory and processing time. MAGIC [3], a language designed for processing medical images, is unusual in being an extension of BASIC.

Languages of the second kind (an example is SUSIE [4]) use segments written originally directly in assembler, with a simple and concise "command" language. They are highly efficient in processing time and memory utilization. SUSIE was written for an uncommon machine (a PDP12), and is therefore unportable, but much image processing is carried out on very few types of machine (e.g. PDP11s), so that there is really considerable scope for "second kind" languages provided the machine type is carefully chosen.

The justification for high-level languages directed specifically towards automated visual inspection is that the widest range of problems within the general field may be investigated with a minimum of effort. A powerful general purpose computer may be used, with operations selected from a library and applied in a desired sequence. The activities being investigated would however be implemented "on line" using special purpose processors which were cheap and fast, but quite inflexible. The high-level languages must be able to imitate these special purpose systems.

All languages discussed share one shortcoming for inspection applications, namely that one standard form of filestructure is provided for the image data. This is the rectangular pixel array (which has to be handled as a sequence of rows or columns in a non-parallel computer), and which is uneconomical and inflexible. SUSIE is somewhat better in having chain and run encoding operations. The purpose of this chapter is to discuss some alternative possibilities for filestructures.

2. FILESTRUCTURE REQUIREMENT

In automated visual inspection the image processing is performed in a sequence of stages (Fig. 1). The first is a CORRECTION stage, in which imperfections in the camera such as non-uniformity are removed. The second is PRE-PROCESSING, in which the information indispensible for the inspection is extracted and retained, and the remaining unnecessary data discarded. The most common pre-processing activity is boundary extraction. From this point onwards, the standard "square array of pixels" format for the data file becomes uneconomical. The main processing is then performed on this reduced data file, which may need to be transferred to the backing store for long-term retention, or be displayed.

An image as sensed directly is a map of the distribution of energy over a two-dimensional surface. It is therefore a real, single-valued, non-negative function of two orthogonal variables. After sampling to reduce it to numerical form for computer analysis, the image comprises a rectangular array of real numbers, generally quantized and scaled to occupy the range 0–255.

Storage in this form is for inspection purposes inefficient and uneconomical. Often, the object being inspected moves along a conveyor belt, so line by line

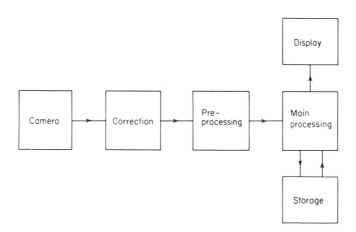

Fig. 1. Stages for image processing in automated visual inspection.

analysis is indicated. The pre-processing stage can often be implemented in real time using special purpose hardware (e.g. by convolving a small number of consecutive scans with a mask). The appropriate form for the file is dictated by the camera used (line scan, random scan), by the processing philosophy (sparse scan, scan pairs), and by the nature of the image (surface of a three-dimensional object), in addition to the task performed. A common form of filestructure capable of accommodating all of these, or at least, all for a surface of given dimension, would be highly desirable. In the next section, some alternative forms are suggested for the data structure.

3. SOME ALTERNATIVE STRUCTURES FOR IMAGE FILES

3.1 Matchsticks

Matchstick images comprise line segments only, though extensions may include other simple shapes such as circles. Experience indicates that the useful information within a scene can often be expressed adequately as a line drawing. A matchstick man, for example (Fig. 2), is readily recognizable as such. Information in scenes is generally contained in discontinuity in tone, texture (and colour if available), and is extracted using gradient finding operators or histogram based level slicing, often followed by operations such as Hough transformation to reduce noise. The discontinuities can then be represented adequately by a matchstick structure. Matchstick images are readily represented by the length, direction and starting point (or, equivalently, co-ordinates

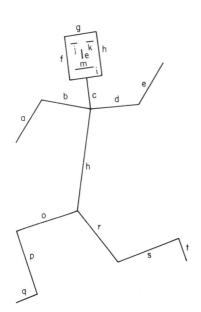

Segment	1st x	1st y	2nd x	2nd y
a	x_a	y_a	x'_a	y'_a
b	x_b	y_b	x'_b	y'_b
c	x_c	y_c	x'_c	y'_c
d	x_d	y_d	x'_d	y'_d
e	x_e	y_e	x'_e	y'_e
f	x_f	y_f	x'_f	y'_f
g	x_g	y_g	x'_g	y'_g
h	x_h	y_h	x'_h	y'_h
i	x_i	y_i	x'_i	y'_i

Fig. 2. Matchstick man and his filestore description.

of starting and finishing points) of the constituent line segments, and provide an economical description.

3.2 Boundaries

A silhouette image is represented completely by its boundary, which is considered to differ from a matchstick on the one hand in having segments always one resolution cell long, and on the other in requiring the ensemble of segments representing the boundary of a real object to obey certain rules. For example, the boundary must be closed, and the elements must form a string, in which each element has two, and only two, other elements as neighbours.

The most famous form for representing boundaries is the well-known Freeman chain code [5]. This [6] can reduce the storage required for a simple shape by two orders of magnitude, with even greater economy if redundancy reducing techniques are used to exploit the property that most boundaries are "smooth". The Freeman chain code is unfortunately sensitive to changes in orientation and location, and no simple algorithm is yet available to compensate for these changes. The contour code obtained by differentiating the chain code [7] is more resistant to change in orientation. Chain code descriptions are easily extracted using consecutive scans with hardware masks, and are easily joined in sections which may be held in serial memory [8].

Another important form for boundary representation is provided by parametric coding [5]. Often, the same methods may be used for representing boundaries as for matchsticks.

3.3 Sparse scan

This approach exploits the property that although all the information within an N by M point representation of an image may be required for an analysis, some operations may be performed on a coarser grid P points by Q, with $P=N/n$ and $Q=M/m$. The consequence is an mn-fold economy in processing time and storage for those operations. For $n=m=10$, this amounts to two orders of magnitude.

Two applications for sparse scan are well established. The registration of images by cross-correlation [9] and the computation of histograms are facilitated by sparse scans, as is the location of isolated "interesting" regions in a field [10] which is largely empty.

3.4 Random scan

This exploits the property that most of the area within the images encountered in automated visual inspection contains no useful information, and should therefore not even be acquired by the camera in the first place. A special form of camera is required for random scan: early users of the technique for track following in nuclear physics [11, 12] used a laser spot scanned by moving mirrors. Subsequently, the dot-scan vidicon camera [13] and an inexpensive image dissector [14] have become available, and provide random scan input in which the scanned spot can be moved under computer control within a field 4096 points by 4096, at low cost. The basic technique involves performing a sparse scan over the whole field until a significant feature (such as a boundary) is encountered. This is then tracked at high resolution, using a search routine which ensures that the boundary is followed irrespective of its changes in direction. This approach is still being developed [15].

3.5 Sequential scan

All cameras except random scan devices acquire image data line by line in sequence. Some cameras (laser scanners and CCD line arrays) can scan only along lines, with the perpendicular movement obtained by the object being examined being moved along a conveyor belt. It is very desirable that as much processing as possible be performed on the minimum number of consecutive lines (ultimately this would be two), so that a minimum of information need be held in store. The scan lines may be stored as analog data, and the processing performed by convolution with a mask.

Recent research [16] has investigated the extent to which processing may be performed using only pairs of scan lines considered in sequence. Although such operations as boundary extraction and chain coding may be so implemented, simpler processing [17] or more accurate measurement [18] may often result if three or more lines are used rather than two. It seems that although many parameters (e.g. area, boundary and location of centroid) may be computed easily using this approach, others such as the lengths of radii from centroid to boundary require the complete boundary to be stored. Thus, although sequential scan cannot itself provide all parameters, it can nevertheless provide a valuable compression of data. The full extent of its capability, and its shortcomings, remain in doubt.

3.6 Three-dimensional surfaces

Special filestructures are required to hold three-dimensional surfaces economically. Representation as a 256-level quantized grey-scale image is highly redundant, and will in any case not describe re-entrant surfaces when the image function becomes multi-valued.

One solution is the extension to three dimensions of the Freeman chain code. This is considered in section 4.2.

4. EFFICIENT DESCRIPTION

4.1 Two-dimensonal filestructure

Producing a universal file for two-dimensional descriptions is not easy, since many of the structures mentioned in Chapter 3 are mutually incompatible. For example, although a Freeman code description of a boundary is easily determined using a sequential scan approach, our experience [19] shows that it is difficult to link the chains describing a complex boundary pattern such as a printed circuit board, despite Batchelor's optimism [4]. The inverse operation, of guiding a search window using a chain encoded description with sequential scan, is horrendous.

The most popular approaches to economical handling of two-dimensional image files use hierarchy. Klinger [20] has examined an approach termed regular decomposition, in which the image is decomposed into successively smaller rectangles, and only those containing useful information retained. This would be useful for handling sparse scans. Unfortunately, most of the filestructures suggested are inherently not hierarchical. There is no reason why one particular segment in a matchstick figure, for example, should necessarily be considered more "important" than another.

4.2 Three-dimensional filestructure

The obvious solution to the efficient representation of three-dimensional surfaces is the extension to three dimensions of the Freeman chain code. To determine this, we imagine the surface to be defined within a three-dimensional "quantization" grid. The surface is further considered to occupy the centre of any cell which it crosses. Each occupied cell lies in the centre of a "block" of 27 cells, within which its immediate neighbours must lie. Like a two-dimensional boundary, a three-dimensional surface must for any real object be closed. Each occupied cell must have exactly eight nearest-neighbour cells which are also in the boundary and hence occupied. Three distances may occur between neighbouring cells, namely, d, $\sqrt{d}$, and $^3\sqrt{d}$, where d is the grid increment.

However, the relationships thus implied in the two directions are not independent. It is probably better to represent the surface by a sequence of two-dimensional chains describing successive "slices" through the surface. The resulting description is complete and unambiguous, and accords with sequential image acquisition.

5. CONCLUDING REMARKS

We have discussed six non-standard data structures useful in image processing applications, particular for automated visual inspection. A recent book [21], and an article by Pavlidis [22] provide a useful review of forms for describing shapes, but much work is required in this field. Since the filestructures discussed are commonly used in on-line applications, they should clearly be provided in any high-level language designed to simulate the on-line activity.

REFERENCES

] Lawden, M. (1977). "The VICAR Image Processing System". Handbook prepared for SRC Appleton Laboratory.

[2] Saxton, J. O. (1978). *"Computer techniques for image processing in electron microscopy"*. Academic Press, London and New York.

[3] Gregory, P. J., Taylor, C. J. and Dixon, R. N. (1977). "Delineation and feature extraction". *SPIE* **130**, 46–52.

[4] Batchelor, B. G., Brumfitt, P. J. and Smith, B. V. D. (1980). "Command languages for interactive image analysis". Proc. IEE **127 E**, pp. 203–218.

[5] Freeman, H. (1961). "On the encoding of arbitrary geometric configurations". *Trans. IRE on Elect. Comp.* **EC–10**, 260–268.

[6] Norton-Wayne, L., Hill, W. J., and Finkelstein, L. (1977). "Image enhancement and pre-processing". *SPIE* **130**, 29–35.

[7] Eccles, M. J., McQueen, M. P. C. and Rosen, D. (1977). "Analysis of the digitised boundaries of planar objects". *Patt. Recog.* **9**, 31–42.

[8] Batchelor, B. G. and Marlow, B. K. (1980). "Fast generation of chain code". *Proc. IEE* **127 E**, 143–147.

[9] Pearson, T. J., Hines, D. C., Golosman, R. S. and Kington, C. D. (1977). "Video rate image correlation processor". *SPIE* **119**, 197–205.

[10] Bretschi, J. and Konig, M. (1980). "Optical and acoustic sensor systems for pattern recognition". Proc. 5th ICAIPC, Stuttgart, pp. 35–50.

[11] Frisch, O. R. (1977). "Sweepnik — a fast, semi-automatic track measuring machine". In *"Machine perception of patterns and pictures"*. Inst. of Phys. Conf. Pub. No. 13, pp. 187–190.

[12] Brooks, C. B., Davey, P. G., Harris, J. F. and Wilkins, C. A. (1972). "PEPR—a random access film scanning system". In *"Machine perception of patterns and pictures"*. Inst. of Phys. Conf. Pub. No. 13, pp. 181–186.

[13] Guildford, L. H. (1978). "Experiments in real-time picture processing". IEE Conf. Pub. No 173, pp. 55–56.

[14] Hamamatsu (1980). "Random Access Camera C1181X for computers". Data leaflet X4-176-178-7-10, Hamamatsu Co.

[15] Mero, L. and Vamos, T. (1976). "Real-time edge detection using local operators". Proc. 3rd. IJCPR, Coronado, USA, pp. 31–36.

[16] Veillon, F. (1979). "One pass computation of morphological and geometrical properties of objects in digital pictures". *Sig. Proc.* **1**, 175–190.

[17] Koulopoulos, C. (in prep.). "Boundary extraction using masks".

[18] Dessimoz, J. D. (1979). "Curve smoothing for improved feature extraction from digital pictures". *Sig. Proc.* **1**, 205–210.

[19] Pullen, A. (in press). "Automated visual inspection of PC boards". Thesis, The City University, London.

[20] Klinger, A. and Dyer, C. R. (1976). "Experiments in picture representation using regular decomposition". *Comp. Graph. and Image Proc.* **5**, 68–105.

[21] Klinger, A., Fu, K. S. and Kunii, T. L. (1977). *"Data structures, computer graphics and pattern recognition"*. Academic Press, London and New York.

[22] Pavlidis, T. (1980). "Algorithms for shape analysis of contours and waveforms". Trans. IEEE on P.A.M.I., **PAMI 1–2**, 301–312.

Chapter Eleven

Overview of the High-level Language for PICAP

B. Gudmundsson

1. INTRODUCTION

At Linköping University a picture processing system, PICAP I, has been operational since 1974 [1]. PICAP I has been used in many applications, especially in projects involving medical images [2]. Algorithm development is the predominant activity in our picture processing laboratory. This makes an interactive system a prerequisite for efficient work in this environment. A high-level language system for PICAP has been developed and implemented. At present a second generation system, PICAP II [3], is being built and the high-level language will be transported to this new system.

2. PICAP I

The main components of the PICAP Picture Processing Laboratory [1] are shown in Fig. 1. TIP is the television I/O-processor. Pictures in PICAP have a fixed size of 64×64 picture elements (pixels), 4 bits/pixel. They can be stored in either one of the twelve hardware registers DR0–DR2, R0–R8 or in the memory of the host computer. The picture processor fetches its operands and stores its results in the register set R0–R8. TIP is controlled by the host computer through a set of parameters, e.g. window positioning, A/D-converter control, camera/monitor selection, sampling density. All picture processor operations generate a number of result parameters that give certain measurements from the resulting picture. The parameters include a histogram of 16 levels, max/min grey-level, etc, and they are automatically transferred to the memory of the host computer when a picture processor operation is completed.

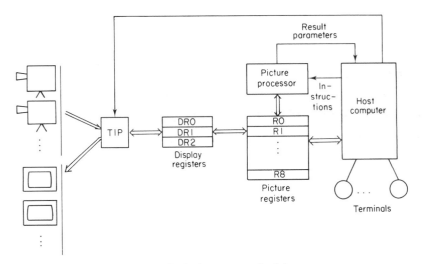

Fig. 1. The PICAP picture processing laboratory.

The picture processor is a parallel processor that transforms one picture into another by means of local 3 × 3 neighbourhood operators. The host computer is a mini of Swedish origin, Datasaab D5/30, equipped with 128 Kbytes of core, disk storage, lineprinter and a number of terminals.

3. PICTURE PROCESSING LANGUAGE

The system is centred around a structured high-level language called PPL (Picture Processing Language) and it contains a monitor, a text editor, an incremental translator for translation of PPL-text into pseudo code, an interpreter and a file system (for procedure library). The system interacts with the user at the level of one line of source code, i.e. as each line is entered, either when a new procedure is being input or when editing is performed, it is syntax-checked and translated into pseudo code. If a syntax error is detected, an appropriate plain English error message is issued and the user is prompted to try again.

When a procedure is completely entered or edited, it is syntactically correct and it has been translated into pseudo code. Execution can follow immediately, i.e. the interpreter can, without further ado, start interpreting the pseudo code. Note that, when in edit mode, lines that are not affected by editing operations are not retranslated since translation is performed on a strictly local basis, line by line.

The mode of system operation described above makes it possible for the user to rapidly enter and modify his procedures and then immediately execute them. The system is written in FORTRAN IV, with the exception of the interpreter which is written in assembly language.

PPL is a structured language in the sense that there is no GOTO; instead a number of compound statements are available to the programmer. PPL can be said to be composed of two distinct parts; one that contains "conventional" language constructs (e.g. arithmetic, control structures) and one where the language constructs directly pertain to picture processing in PICAP, i.e. they reflect the architecture shown in Fig. 1.

In the remainder of this section, some of the features of PPL will be described. No complete formal description of language syntax will be given; instead we shall resort to more informal discussions that will hopefully convey the highlights of PPL to the reader.

3.1 Procedures

<procedure>::=<heading><decl .part><statement part>

Procedures are separately entered (and thus separately translated). This makes it possible for the user to build a subroutine library.

The procedure heading gives the procedure a name, it names the formal parameters (if any) and it specifies whether the procedure is to return a value or not (i.e. if it is a function). Heading examples are:

PROGRAM TEST
PROGRAM THRESH (UPPER, LOWER)
FUNCTION MAX (LIMIT):INT (integer function)
PROGRAM F (ALPHA, *BETA)

An asterisk preceding a formal parameter name indicates that the parameter is to be passed by reference (instead of by value).

3.2 Declaration part

The datatypes of PPL are the following:
integer
real
boolean
array (max 3 subscripts)
binary picture (1 bit/pixel)

picture (8 bits/pixel)
template element (element in 3 × 3 picture operator, see section 3.3).
All programmer defined identifiers must be declared. To save the user some
work, identifiers, I,J, . . .,N are standard declared integers (with principal use
as loop control variables). Variables of type PICT are considered two-
dimensional integer arrays. They are stored in a packed format, and automatic
packing/unpacking is done when storing/fetching picture elements.
Declaration examples are:

> INT UPPER, LOWER
> ARRAY OF BOOL FLAGS (16,2), SET(10)
> PROC THRESH, RESET (procedures)
> IFCN MAX, MIN (integer functions)
> BPICT MASK1 (128,128)
> PICT MASK2 (1,96), START (10,10)

It should be noted that, due to hardware constraints (Fig. 1), the present
implementation of PPL supports only pictures of a fixed size, namely 64 × 64,
4 bits/pixel. Thus, in the present implementation, a variable declared PICT is
a picture in this standard format and there is no need to specify picture
dimensions.

3.3 Picture operators

The PICAP picture operators [1] operate in parallel over all 3 × 3 neighbourhoods of the picture, i.e. the new value to be assigned to a pixel is dependent on
the values of its eight closest neighbours. There are two kinds of operators;
arithmetical and logical. Let's look at a pixel q_0, and its eight neighbours:

$$q_4 \ q_3 \ q_2$$
$$q_5 \ q_0 \ q_1$$
$$q_6 \ q_7 \ q_8$$

An arithmetical operator is depicted

$$\omega_4 \ \omega_3 \ \omega_2$$
$$\omega_5 \ \omega_0 \ \omega_1 \ N$$
$$\omega_6 \ \omega_7 \ \omega_8$$

where ω_i and N are signed integers. When the operator is applied to q_0 we get
the weighted sum (convolution)

$$q_0^+ = 2^{-N} . \sum_{i=0}^{8} \omega_i . q_i$$

where q_0^+ designates the new value of q_0.

A logical operator is also depicted by a 3×3 matrix:

$$\begin{matrix} m_4 & m_3 & m_2 \\ m_5 & m_0 & m_1 \\ m_6 & m_7 & m_8 \end{matrix} \; T$$

$$\text{where } m_i = \begin{cases} X & (\text{'don't care'}) \\ \text{or} \\ R_i c_i \end{cases}$$

R_i is a relation (equal, less than, greater than), T and c_i are unsigned integers. If for all i, q_i matches m_i then $q_0^+ = T$, else $q_0^+ = q_0$.

In PPL the operators are depicted as 3×3 matrices and they are declared in the declaration part; logical operators under the heading TEMP and arithmetical under MATR.

Examples:

```
TEMP EDGE(R8), TH
X  X  0              X  X  X
4  1  1  2           X  E1 X  0
A  A  1              2  >5 2

MATR LAPLAC, GRAD
0   1  0             1  0  -1
1  -4  1  0          G1 0  G2 2
0   1  0             1  0  -1
```

Two logical operators (EDGE and TH) and two arithmetical operators (LAPLAC and GRAD) are declared. In the logical operators a plain integer implies the relation "equality". Operators may be followed by a list of modifiers in parentheses (e.g. R8 means 8-rotation). In the logical operators, identifiers A and E1 are variables of type template element (see section 3.2). These variables can take on values of template expressions:

$$\langle\text{temp .expr .}\rangle ::= \text{'X'} | [\text{'>'}|\text{'}|\text{'}\text{'<'}] \langle\text{integer .expr .}\rangle$$

where 'X' designates "don't care".

The variables in the arithmetical operators (G1,G2) are considered to be of type integer.

Allowing picture operator elements to be variables makes it possible to redefine operators during execution; a feature that is very useful when developing algorithms.

The picture processor allows for chaining of logical operators (maximum 8) so that if the first operator does not match, the next operator in the chain is tried, and so on. Operator chains are specified in the declaration part. Example:

```
:
TEMP T1, T2
X  X  X          X  X  X
X  >6 X  15       X  X  X  0
X  X  X           X  X  X
CHAIN THRESH: T1*T2
:
```

When THRESH is applied to a picture all pixels with values >6 will be set to 15, all others will be set to 0.

3.4 Compound statements

The control structure of PPL is based on a number of iterative and conditional compound statements.

Excerpts from the grammar:

<statement part>::='BEGIN'<computation>'END'
<computation>::={<statement>}
<statement>::=<simple statement>|<compound>
<compound>::=<do>|<while>|<for>|<if>|<case>
<do>::='DO'<computation>'OD'
<while>::='WHILE'<bool.expr.><do>
<for>::='FOR'<assignm.>'TO'<integer.expr.><do>
<if>::='IF'<bool.expr>'THEN'<computation>
['ELSE'<computation>]'FI'
<case>::='CASE'<integer.expr.>'OF'<integer.const.>
{<label>':'<computation>}
['CELSE'<computation>]'ESAC'
<label>::=['−']<integer.const.>

The DO–OD compound has the semantics of "do forever". The use of Algol 68 style explicit closing delimiters (OD, FI and ESAC) solves the dangling ELSE and analogous problems and eliminates the need for DO–END or BEGIN–END to delimit compound statements. PPL has no GOTO statement but a mechanism for loop escape is provided in the form of a simple statement LEAVE, which is a forward jump out of one level of enclosing DO–OD. Another simple statement, TURN, is a forward jump to the nearest closing delimiter OD.

3.5 Picture processing

This part of PPL reflects the architecture of PICAP (Fig. 1). It includes declaration of the picture operators to be used in the procedure and a number of simple statements for operations on pictures, picture transfers, picture I/O, etc. The hardware picture registers are indicated by keywords, R0–R8, DR0–DR2, and operations upon their contents are expressed in functional notation, e.g.

$$R3 = EDGE(R1)$$

where EDGE is the name of an operator that has been specified in the declaration part.

Transfers of pictures between registers and host computer memory take the form of assignments, e.g.

$$R0 \ = R2$$
$$DR1 = R3$$
$$R1 \ = MASK$$

where MASK is a variable of type picture (i.e. a picture stored in host machine memory).

As was mentioned earlier, each picture processor operation generates a number of result parameters that pertain to certain measurements in the resulting picture. The parameters are automatically transferred to an area in memory when the operation is completed. In PPL these parameters are reached through a number of standard declared read-only integer and array variables, e.g. a one-dimensional array called HIST contains the current histogram.

4. EXAMPLE TERMINAL SESSION

Some features of system operation will now be described by means of an example terminal session. A simple procedure to test some varieties of a symmetric arithmetic picture operator is entered and executed. The picture to be operated upon is assumed to reside in DR1 (Fig. 1). For clarity, characters output by the system are indicated by bold type.

//ENT
0001   PROGRAM TESTSY
0002   MATR SYM
0003   −1 P −1
0004   P Q P !

```
****        ILLEGAL CHARACTER:! ****
0004        P Q    P  1
0005        −1  P  −1
0006        BEGIN
0007        RO=DR1                    #PICTURE TRANSFER
0008        READ(Q) READ(LIMIT)
****        LIMIT NOT DECLARED ****
0008        **DEC
0006        INT LIMIT
0007        $
0009        READ(Q) READ(LIMIT)
0010        FOR P=−2 TO 2 DO
0011            R1=SYM(RO)          #PICAP OPERATION
0012            IF GMAX−GMIN>LIMIT THEN
0013            PRINT(P:5X,'P=',1) PHIST FI
0014        OD
0015        END
0016        **OUT
//RUN
  ?
  4
  ?
  6
        P=0
HISTOGRAM:
```

0.	1.	2.	3.	4.	5.	6.	7.	8.	9.	10.	11.	12.	13.	14.	15.
2224	518	137	44	12	6	4	2	7	9	19	30	45	81	238	720

```
//EDT       TESTSY
**APP       12
*0012           IF GMAX−GMIN>LIMIT THEN
0013            DR2:R1
            **** INCORRECT STATEMENT ****
0013            DR2=R1        #DISPLAY RESULT
0014        $
**LIS       1,100
*0001       PROGRAM TESTSY
*0002       MATR SYM
*0003       −1  P  −1
*0004        P Q    P  1
*0005       −1  P  −1
*0006       INT LIMIT
*0007       BEGIN
```

```
*0008      RO=DR1               #PICTURE TRANSFER
*0009      READ(Q) READ(LIMIT)
*0010      FOR P=-2 TO 2 DO
*0011          R1=SYM(RO)       #PICAP OPERATION
*0012          IF GMAX-GMIN>LIMIT THEN
*0013          DR2=R1           #DISPLAY RESULT
*0014          PRINT(P:5X,'P=',1) PHIST FI
*0015      OD
*0016      END
**OUT
           SAVE(YE OR NO)
NO
//RUD
```

The user enters the input mode by means of ENT and he is prompted by consecutive line numbers. The operator to be tested (SYM) contains two variable elements (P and Q). The user forgot to declare LIMIT and therefore he types the command DEC, which, as was mentioned earlier, allows him to append lines to the declaration part. Input according to DEC, as well as APPEND and INSERT, is discontinued when the user types "$". GMAX and GMIN (max/min grey-level) are standard declared integer variables that pertain to the result of the latest picture operation. PRINT is a statement for formatted printing and PHIST prints the current histogram. The input mode is left by means of command OUT, and the procedure can be executed (RUN). Now the user wants to modify the procedure so that the resultant pictures that pass the test in line 12 will be displayed. This is done by transferring the picture to one of the display registers (Fig. 1). Edit mode is entered (EDT) and a new line is appended to line 12. When this is done the user gives command LIST and gets a listing on the terminal (command TYP lists on the line-printer). Edit mode is left (OUT) and the procedure can be executed. This time the user gives command RUD (run and display), which means that execution will pause each time a display register has been loaded. A more general execution pause capability is provided by the simple PPL-statement PAUSE.

5. CONCLUDING REMARKS

PPL has been implemented and is now running in our picture processing laboratory. As was mentioned earlier, a new picture processing system called PICAP II [3] is being built and PPL will be transported to this new system. Since the architecture of PICAP II differs from that of PICAP I some changes in the parts of PPL that pertain to picture processing will be necessary.

B. Gudmundsson

REFERENCES

[1] Kruse, B. (1976). "The PICAP picture processing laboratory". Proc. 3rd IJCPR, Coronado, USA, pp. 875–881.
[2] Rudgård, A. and Nordell, L-E. (1979). "Image analysis of parathyroidea glands". Proc. 5th Int. Conf. on Stereology, Salzburg.
[3] Kruse, B., Danielsson, P-E. and Gudmundsson, B. (1980). "From PICAP I to PICAP II". To appear in *Special computer architectures for pattern processing*, Fu, K. S. and Ichikawa, T. (eds). CRC Press Inc., Cleveland.

Chapter Twelve

Comparing some High-level Languages for Image Processing*

A. Maggiolo-Schettini

The discussion at the workshop was centred upon what an image processing language should be. For many people an image processing language turned out to be simply a set of FORTRAN subroutines (like PAX and SLIP) and the design and the circulation of standardized libraries was advocated. Preston [6] reviewed something like thirty languages consisting of sets of FORTRAN subroutines which are accessed by a standard call, and languages which totally control the computer environment acting as their own operating system, interpreter or compiler and containing full file management capability (like GLOL). (I would prefer to call these systems rather than languages.) Preston concluded that, in his opinion, the trend was in the direction of utilizing the existing operating system of the computer and loading the image processing language as a library of subroutines which are accessed via a command decoder. The categories of functions performed by these languages are: (1) utilities; (2) image display; (3) arithmetic operation; (4) geometric manipulation; (5) image enhancement; (6) image analysis; and (7) decision theoretic. Very few of the languages cover all the categories. Note also that only some of them are designed for real or emulated parallel processing.

Levialdi [3], in his review, quoted a number of efforts in developing high level languages for parallel architectures like those of ILLIAC IV, STARAN and PEPE. The languages GLYPNIR, STARAN HLL and PFOR are based on ALGOL (the former two) and FORTRAN (the latter) and besides high-level control structures they also have instructions which make a precise reference to

---

* This chapter has been written almost one year after the Windsor Workshop on High-level Languages for Image Processing and contains some considerations on the workshop and on results which have appeared or have become known to the author afterwards.

the processing elements of the machine and allow, for instance, the routing of the results from one processing unit into another. Note that the above mentioned machines and languages are not designed in particular for image processing.

Along the same line, seems to be Kruse's proposal of the high-level language PPL for his pipeline machine PICAP (note that both the machine and the language have been designed for image processing). The language PPL has high-level control constructs and allows the definition of logical and arithmetical templates to operate on a picture, but it also allows referencing the special registers that contain information stored by the PICAP processor.

The favour of the workshop among attendees was gained by the idea of designing programming languages for image processing which would be high level, i.e. machine independent and therefore transportable; and capable of expressing the needs of people working in different branches of image processing and therefore adaptable. It seemed also to be rather reasonable to use as a kernel of the language some of the existing high-level languages defined mostly for numerical and sequential computation. This was for two main reasons: first in order to exploit the experience gained in defining and implementing such languages over more than twenty years; and secondly in order to experiment with parallel processing of images expressed with parallel constructs which may also be implemented on sequential machines (like those that most people have at their disposal) in terms of the existing constructs for sequential computations. It turned out that at least four proposals followed this line: two by Uhr and Douglass of languages based on PASCAL [9, 1], one by Kulpa of a language based on ALGOL 68 [2] and one by Levialdi and others (among them, the author of this chapter) of a language based on ALGOL 60 [4]. At the time of the workshop only the first and the fourth proposals contained an implementation of some constructs and some examples could be shown.

The discussion concentrated on the language that should have been chosen as a base for the image processing language rather than discussing the merits of the proposed constructs. The suggestion of PASCAL was almost unanimous mostly because it was believed to be also available on small computers, which is not the case for ALGOL 60 and ALGOL 68.

The purpose of this chapter is to review the above mentioned proposals and also two new ones, which appeared during the year following the workshop, from the point of view of the proposed constructs [see 7, 8]. We shall briefly speak also of ACTUS [see 5] which is a new high-level language for parallel processing (not for images in particular) which is based on PASCAL and is translated into the code for ILLIAC IV by an auxiliary computer to ILLIAC IV.

We shall not describe the languages and their constructs systematically, but we shall concentrate our attention on a number of features we believe to be crucial to measure the expressive power of a language for image processing. We shall give examples of statements from which the reader should understand similarities and differences of the solution offered, if any, by a language to a certain problem. In many cases the reader is asked to use his intuition for understanding the meaning of a certain statement from its syntax.

The features we will consider are the:

(a) possibility of defining arrays on which to operate in parallel;
(b) possibility of selecting a subarray of an array for partial processing;
(c) possibility of comparison between the neighbourhood of an element of an array and a given pattern;
(d) availability of parallel instructions with global control;
(e) availability of parallel instructions with local control.

(a) In Parallel PASCAL 1 and Pascal PL, arrays on which one wants to operate in parallel have a special definition (in the former it is specified that the compiler should provide assignment of the elements of the array to the PE's starting from the innermost dimension). The forms are as in the respective examples:

a: PARALLEL ARRAY $[1..5, 1..10, 1..12]$ OF integer
‖**dim** & $[0..127, 0..127]$

(in Pascal PL, & and * are used for boolean and integer respectively and the defined dimensions hold for all the arrays, as the dimensions of the arrays must be compatible to operate correctly).

PAL provides a definition of ROW and PICTURE. In L one can define groups of images, every image having one or more bands, every band being a rectangular array of pixels. The types may be integer (not specified) and boolean. One may declare how many bits should be reserved for each pixel. A default size for images may be defined. Some examples follow:

IMAGE GROUP : $c(3) <100,200>$

meaning that the group has three images of size $<100,200>$,

IMAGE B[3] $<10,20>$

meaning that B has three bands with size $<10,20>$,

DEFAULT IMAGE : $<1024,520>$
BITS PER PIXEL : 8.

PIXAL has the standard array declaration of ALGOL 60 with the additional types *binary* and *grey*. Douglass' PASCAL Extension has standard PASCAL array declaration.

Therefore only Parallel PASCAL 1 and Parallel PL require that arrays on which one may want to operate in parallel have a special declaration.

Only Pascal PL and PIXAL have provisions for constraining the image data to be embedded in a chosen background. Pascal PL has the construct

||**border**: = bordertype

where bordertype may be 0 (for "false"), or 1 (for "true"), or 2 (for "what is contained by the nearest cell within the array"). The default condition is border equal 2.

PIXAL has the construct

edge-of array identifier **is** arithmetic expression

which allows the image data to be embedded in a background whose value is given by the expression. The default option is 0.

All the above languages allow parallel assignment to arrays of the result of operations on arrays. In Parallel PASCAL 1, L and PIXAL it is of the form

$$A: = B + C$$

In Pascal PL it is of the form

||**set** $A: = B + C$

PAL and PASCAL Extension require the specifications of dummy indices. In PASCAL Extension the construct is slightly more complicated because there is the insruction SCAN to let the indices vary from the right to the left and the instruction SPLIT explicitly to express the parallelization of the operation splitting the task among the available processors. An example follows:

SPLIT I: = 1 TO N SCAN A DO SUM: = SUM + A[., .]

(b) In Parallel PASCAL 1 it is possible to select entire dimensions of an array, like in

$$A[,j]$$

to select the jth column of A, or intervals along a dimension

$$A[1 . . 10].$$

Besides, it is possible to apply operators along one specified dimension only giving, as a result, one array with only that dimension

reduce(array,dimension,operator)

or expand an array in a new dimension

expand(array,dimension,size)

obtaining as a result the new dimension inserted at axis "dimension" and the data in the array is repeated "size" times along this dimension. It is not clear if in Pascal PL selecting a subarray is possible.

In L there is the possibility of positioning and viewing a rectangular portion of an image by the concept of window. The window size must be defined

$$W:=<20,10>,$$

it can be varied during the program

$$W:=W+<1,2>,$$

must be positioned on the wanted array

$$POSITION (W,A,X,Y)$$

and finally it can be used as in the examples

$$C:=A:W.$$

In PAL there are similar facilities. In PIXAL a frame can be defined giving the size in the various dimensions, the centre (if not the geometric centre) and the wanted subarray can be found and operated upon. For instance

$$overlap(F,A)$$

returns an array with the same dimensions and size of F and whose elements are the same as those in the environment of the current element of A (see also later).

In PASCAL Extension the concept of window is similar to the one in PIXAL. In the example

$$SCAN\ WINDOW\ A\ FROM\ I,\ 1\ TO\ 5,\ N\ DO\ SUM:=SUM+A[\ .,\ .]$$

the first five rows of the $N \times N$ array A are summed up.

(c) In Parallel PASCAL 1 there is no definition of masks. (The comparison can certainly be expressed by testing elements, but it is less immediate).

In Pascal PL, masking can be realized by compounds which express the conditions under which the elements of the neighbourhood contribute to the value of the considered element, as in

$$\|\mathbf{set}\ a:=b[+(0:-1=1,0:0=0,0:1=1)].$$

In L it is possible to define boolean images (having only 0 and 1, as values of the elements) for the purpose of masking.

In PIXAL, masks are specified by dimension size in the various dimensions and a centre for proper positioning (optional). To masks, values may be assigned and can be used as in the example:

compare(M,A)

which gives as a result, true, if the subarray of A with the same dimensions as M and centred over the current element of A (see later) is equal to M, otherwise false is given as a result. Note that, thanks to a special symbol, PIXAL masks do not need to be rectangular and also the positioning is left to the programmer. PASCAL Extension does not provide a similar construct.

(d) This means the possibility of controlling parallel operations by a global test on an array, i.e. testing all the pixels for a given predicate. It is possible in PIXAL. Take as an example

if $A \neq 0$ **then** A:=1

It is not available in the other languages we have considered but it is possible in ACTUS (see later).

(e) The diameter size of the predicate which has been computed is limited to a nearest neighbourhood of the considered element.
In Parallel PASCAL 1 it is expressed like in the example

IF B>C THEN A:=B+C

which has the meaning that the assignment is done (simultaneously) only for those elements for which the test gives true.
An example in Pascal PL is the following

∥**if** featurei[+(0:1,0:−1)]* featurej[+(1:0,−1:0)]>11
∥**then** labeli+19
∥**else** labell*20

which has the meaning that, in parallel, the elements of labeli for which the test gives true are incremented by 19 and the elements of labell for which the test gives false are multiplied by 20.
An example in L is the following

FOR ALL <i,j> IN A DO IF A<i,j>=0 THEN A<i,j>:=1
ELSE A<i,j>:=0

The construct FOR ALL in PAL is quite similar.
In PIXAL, parallel instructions with local control are implemented by the construct par-parend.
For example

par if $A=0$ **then** A:=1 **else** A:=0 **parend**

In Parallel Extension it would be

SPLIT ON A DO IF A[.,.]=0 THEN A[.,.]:=1
ELSE A[.,.]:=0

Let us turn now to the choice of the support language and stage of development of the implementation.

It was already said that the supporting language is PASCAL in three cases, ALGOL 60 in two and ALGOL 68 in one. It should be said that in the case of L and PIXAL the authors claim that their choice was due to the availability of ALGOL on the computers at their disposal. In L it is possible to associate to an array attributes which reminds characteristics of languages with a data structure more rich than that of ALGOL 60. On the other hand Reeves [8] mentions the lack of dynamic arrays as a serious limitation of PASCAL.

Parallel PASCAL, L, PAL and Parallel Extensions are at the stage of proposals. Pascal PL, and PIXAL have experimental implementations on sequential machines that do already work.

We want now to remark that ACTUS, which was not designed for image processing in particular, has many constructs similar to those of the languages discussed above.

Arrays to be used in parallel instructions may be defined indicating the desired degree of parallelism. It is also allowed to work on a subarray of the entire array. See the following example.

> **var** parb:**array**[1:100] of integer;
> **index** first 50=1:50;
> parb[first 50]=parb[first 50]+parb[first 50 **rotate**−1]

This means that of the one hundred elements of parb the first 50 are considered and elements from 2 to 50 are added to the elements from 1 to 49, and element 1 is added to element 50 in parallel.

There are parallel instructions with global control (like in PIXAL). See the following (self-explaining) examples

> **while any** (a[1:50]<b[1:50]) **do** a[1:50]:=a[1:50]+1
> **if any** (a[40:70]>0) **then** a[40:70]:=a[40:70]−1
> **if all** (a[2:79]>b[2:29]) **then** b[2:29]) := 0
> **else** a[2:29]:=0

There are also parallel instructions with local control, for example

> **while** a[1:50]<b[1:50] **do** a[#]:=a[#]+1

A precompiler of ACTUS constructs, which are not in PASCAL, should be constructed shortly. The compiler PASCAL-P working on a computer auxiliary of the ILLIAC IV and which translates in ILLIAC IV code has been built.

ACKNOWLEDGEMENTS

The author thanks Bob Douglass, Adolfo Guzmán, Zenon Kulpa, Stefano Levialdi, Tony Reeves, Len Uhr, Alan Wood for useful information and helpful discussions.

REFERENCES

[1] Douglass, R. J. (1979). "Extension to PASCAL for image processing". (Preliminary report.)
[2] Kulpa, Z. (1981). "PICASSO, PICASSO–SHOW and PAL". (This volume.)
[3] Levialdi, S., Maggiolo-Schettini, A., Napoli, M. and Uccella, G. (1979). "Considerations on parallel machines and their languages". Presented at the Windsor Workshop.
[4] Levialdi, S., Maggiolo-Schettini, A., Napoli, M., Tortora, G. and Uccella, G. (1979). "Preliminary results on the construction of an image processing language (PIXAL)". Presented at the Windsor Workshop.
[5] Perrot, R. H. (1979). "A language for array and vector processors". ACM Trans. on Prog. Lang. and Syst., pp. 177–195.
[6] Preston Jr., K. (1979). "Image manipulative languages: a preliminary survey". Presented at the Windsor Workshop.
[7] Radhakrishnan, T., Barrera, R., Guzmán, A. and Jinich, A. (1979). "Design of a high level language for image processing". Tech. Rep. PR–78–22, IIMAS, Nat. Univ. of Mexico.
[8] Reeves, A. P. (1979). "Parallel PASCAL: a provisional description". (Preliminary Report.)
[9] Uhr, L. (1979). "A language for parallel processing of arrays, embedded in PASCAL". Tech. Rep. 65, Comp. Sci. Dept., Univ. of Wisconsin-Madison.

Chapter Thirteen

An Image Analyser

C. Lantuejoul

1. INTRODUCTION

The prototype of an image analyser (IA) designed and built by the Centre du Morphologie Mathematique in 1975 is presented in this chapter. The function of this instrument is to help the specialists in the natural sciences (biologists, geologists, metallurgists, etc.) to extract quantitative information from the images of their specimens (slides, polished sections, etc.).

Thus, the IA performs measurements. Unfortunately, it turns out in practice that the images are never ready for direct measurements. They must be transformed into other images, which are "measureable". As a consequence, the IA also performs image transformations. In short, image transformations are only a necessary step. Measurements are the final aim.

In order to be significant, the measurements must be scarcely altered by a slight modification in the original image. This necessarily requires that the image transformation must satisfy certain properties of continuity [1, 2]. Accordingly, any image transformation is not necessarily adequate to perform quantitative image analysis. However, the class of possible transformations is rather large. For instance, it can be shown that any image transformation which is an iteration of local transformations [3, 4] belongs to this class [2]. They are not the only ones.

The IA is an instrument devoted to quantitative image analysis. This is a special-purpose computer, in which the basic data-elements are binary images. Binary image transformations and measurements are the two main operations which can be performed on these data elements. In order to make the IA easily usable, an interpretation language has been developed with which it can be programmed.

In so short a chapter, it is impossible to give a comprehensive description of the IA and of its language. We present here only their salient features, and illustrate them with two commented programs.

The reader who would like to have further information, may consult Klein [5] for the hardware, and Digabel [6] for the software.

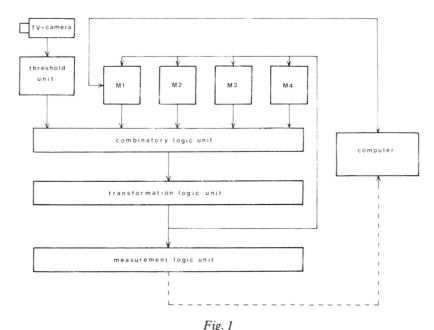

Fig. 1

2. THE HARDWARE

Figure 1 shows the image flow diagram of the IA. The objects to be analysed (photographs, slides or polished sections, etc.) are scanned using a TV camera. The resulting analog signal is thresholded with respect to the light intensity, and sampled at a frequency of 5 MHz to produce a binary image defined on a hexagonal raster.

Four memory planes of size 256 × 256 have been set up to store binary images. A visual display system allows the user to view at any time the original grey-tone image, or the threshold image, or the contents of one or several of the memory planes.

The inputs of the combinatory logic unit are the threshold image and the contents of the four memory planes. The output of this unit is a binary image which is a boolean combination of the input images.

The output image of the combinatory logic unit is the input of the transformation logic unit. This unit transforms a binary image into another binary image. Two main transformations are hardwired in this unit:

(i) With the *first point transformation*, at most one pixel is set to 1 in the output image. This pixel is the upper left pixel set to 1 of the input image. The first point transformation is used as a blob detector in an individual analysis of blobs (see, for example, the commented programs of section 4).

(ii) *The neighbourhood transformation* admits a parameter which is a family of binary hexagonal templates $V_1, V_2, \ldots, V_n$. A pixel is set to 1 in the output image if, and only if, the hexagonal neighbourhood of the pixel-values around it matches one of the V_is. Let us give some examples:

(a) The template

is the parameter of an hexagonal erosion.

(b) The family of templates such as

where a dot stands either for 0 or 1, characterizes a linear erosion.

(c) If the templates are

up to a rotation by a multiple of $\pi/3$, the neighbourhood transformation is a contour detector.

The duration of such a transformation is 20 ms. The output image of the transformation logic unit can be stored into one or several of the four memory planes M1, ..., M4. It can also be used as input of the measurement logic unit. The output of this unit is a numerical value characteristic either of the image's geometrical properties (surface area, perimeter, etc. . .) or of its topological properties (connectivity number, etc.). This numerical value is automatically sent into a mini-computer.

The functions of the mini-computer are threefold:

(i) Numerical calculations can be performed on the data provided by the measurement logic unit.

(ii) The mini-computer is equipped with a system of floppy disks, in which binary images can be stored. The memory plane M1 has been specially hardwired to have a direct connection with the mini-computer. The transmission time of an image is rather slow (about 2 s). This system is used whenever an algorithm requires more than four images to be preserved at the same time. It can also be used to create a library of images, although users rarely do so.

(iii) The mini-computer is used as an intermediary between the user and the IA. It controls the different units of the IA, and the language interpreter is implemented on it.

The IA has some other hardware accessories that we do not describe here. Let us only mention that it is possible to write into the memory planes either with a joystick, or by specifying the co-ordinates of the pixel to be modified. The joystick is an invaluable tool to eliminate obvious artefacts stemming from a poor threshold. Pixel modification is used to produce realizations of random sets which are particularly adapted to the quantitative description of images with irregular or complex structures [7, 8, 2].

3. GENERAL DESCRIPTION OF THE SOFTWARE

A high-level interpreted language, MORPHAL (morphological algorithmic language), has been created in order to facilitate the use of the IA by the users. The MORPHAL statements are of two different types: the first type consists of arithmetic, branching, and input–output statements. They are very similar to those encountered in languages such as Basic or Fortran, and do not warrant any further explanation. In the following, attention will be focused on the second type which refers to image analysis.

The language has been designed so as to emphasize the basic concepts of quantitative image analysis, the image transformations and the measurements. The syntax of the statements has the general following form:

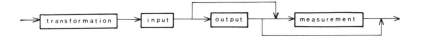

3.1 Transformation argument

This argument contains the mnemonic code of the transformation and its parameters, if necessary.

The mnemonic codes are FPO for the first point transformation and RNG ("ring-logic") for the neighbourhood transformation.

The parameter of the neighbourhood transformation is a family of binary hexagonal templates. A template is given by the value of its centre and of the points of its contour, starting with the right lower point, and turning clockwise. The possible values are:

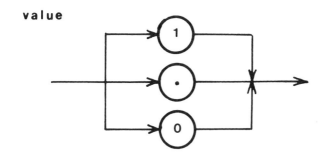

where the dot means "1 or 0". With the IA, the neighbourhood transformation only admits an elementary family of neighbourhoods (efn). The efns are defined by:

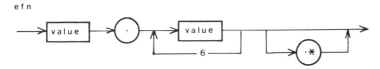

where the symbol * signifies a set of templates consisting of the given template rotated by multiples of $\pi/3$. For instance, the parameter of the contour detector given in the last paragraph is written $1, \ldots \ldots 0, *$.
More generally any family of binary hexagonal templates could be given by:

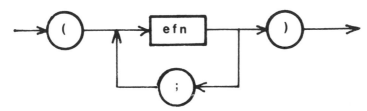

For instance, using this notation, the detector of triple points in a skeleton is $(1,101010,*;1,1\ldots\ldots1,*)$.
In practice, several specific neighbourhood transformations are very often encountered. Rather than specifying the parameter every time, they have been assigned a specific mnemonic code. These transformations are:

NOL (no logic transformation)   RNG $(1, \ldots \ldots)$

ERO (erosion)   RNG (1,111111)
DIL (dilation)   RNG (1,......;.,1.....,*)

3.2 Input argument

This argument specifies the image which is to be input to the transformation logic unit. This image is a boolean combination of the contents of the four memory planes M1, ..., M4, and of the threshold image TH:

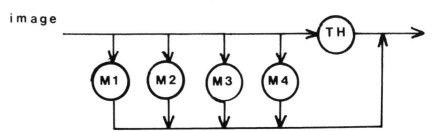

Due to technological constraints of the IA, any boolean combination of the five images is not necessarily allowed. The syntax of the allowed combinations is given by:

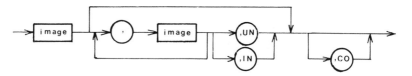

where the UN, IN and CO mean respectively union, intersection and complementation. For instance, $\overline{M1 \cap M2}$ is an admissible combination, whereas $\overline{M1} \cap \overline{M2}$ is not. The difference between them is that the intersection operation is made between two images in the first call; and between two boolean combinations of images in the second case. If a more general hardware accepted any boolean combination, then a possible syntax of the input argument can be extended by using the double recursive formula:

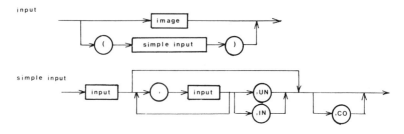

Using this description, the image $\overline{M1 \cap M2}$ is denoted by (M1, M2, IN, C0), whereas ((M1, C0), (M2, C0), IN) stands for $\overline{M1} \cap M2$.

3.3 Output argument

This argument specifies the memory planes in which the output image is to be stored. Its syntax is very simple:

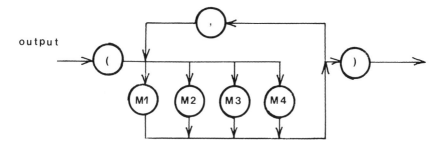

For instance the output argument (M1, M3) means that the transformed image must be stored into the memory planes M1 and M3.

3.4 Measurement argument

This argument specifies the type of the measurement to be performed on the transformed image, and the name of the variable in which the measurement is stored. Its syntax is:

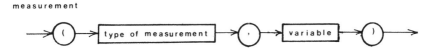

There exist 3 main types of measurements
AR—surface area;
PR—perimeter;
NC—connectivity number.

4. TWO COMMENTED PROGRAMS

4.1 Elimination of artefacts in a population of cells

We start with two images in the two memory planes M1 and M2. M1 contains the image of a population of cells, consisting of isolated cells, clumps of cells, and artefacts. M2 contains the image of a population of nuclei. We want to produce an image, consisting of only the isolated cells (cf. Fig. 2).

cells   nuclei   isolated cells ?

Fig. 2

To do this, we are going to use the following criterion: an isolated cell has one and only one nucleus.

```
*
* detection of isolated cells
*
```

nol (m3,(m3,co),in)(m3,m4)   (1)
visu (m1,m2,m3,m4)   (2)

```
* reconstruction in m3 of a blob of m1
```

1   fₚₒ (m1)(m3)(ar,j)
10   i=j
      dil (m3)(m3)   (3)
      nol (m1,m3,in)(m3)(ar,j)
      if (i .eQ .j)20,10

```
* is this blob an isolated cell?
```

20   nol (m2,m3,in)(nc,n)   (4)
      if (n .eQ .1)30,40

30   nol (m3,m4,un)(m4)   (5)

* towards the analysis of the next blob

40 nol (m1,(m3,co),in)(m1)(ar,k)
    if (k .eq .0)100,1
/*

4.1.1 Comments

(1) The contents of M3 and M4 are cleared;
(2) From here on, the contents of the four memory planes are displayed (visualized);
(3) The reconstitution of blobs of M1 is made by successive dilations, using an iterative algorithm [9]. This reconstruction method is far from being optimal. A better method is given in the next example;
(4) Counting the number of nuclei hitting the reconstructed blob;
(5) At the end of the program, the isolated cells are in M4.

Of course, an individual analysis is particularly tedious when M1 contains numerous blobs. The reader will find in Lantuejoul and Beucher [10] another algorithm allowing all of the blobs of M1 to be analysed at the same time.

4.2 Computation of the edge numbers of grains in a tessellation

The structure presented in Fig. 3 is very common in metallurgy. This is an aggregate of non-overlapping grains. The grains of the aggregate are curvilinear polygons and, due to the surface tension forces on them, their vertices are triple.

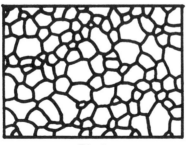

Fig. 3

In order to compute the edge number of a grain within a population, we can imagine the following procedure:
(i) the grain is selected within the population (cf. Fig. 4.1)
(ii) the grain is dilated, in order to hit all its neighbouring grains (cf. Fig. 4.2)

(iii) the dilated grain is then intersected with the initial population (cf. Fig. 4.3). It splits into several parts, and the number of these parts must be equal to the edge number of the grain plus one.

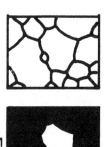

 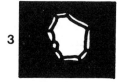

Fig. 4

The important matter is to choose the size of the dilation. This size must be large enough for the dilated grain to hit its neighbours as often as necessary, and small enough so as not to hit the second neighbours. If the domain between grains has not a very homogeneous width (which is almost always the case), these two conditions are incompatible and no value can be found. If the width of the domain between grains is homogeneous, the procedure is effective. Therefore, before applying the edge number procedure, the domain between grains must be homogenized. This is done by a skeletonization.

```
*
* computation of the edge number of grains
*


    dim n(20)
    visu (ca)
    halt                            (1)


* at this point, the user thresholds the grains in white
    nol (th)(m1)
    visu (m1)
    halt


* at this point, the user can eliminate the artefacts
* of the image contained in m1 by using the joy-stick

* skeletonization                  (2)
```

```
      j=0
10    i=j
      rng (1,......;0,11 .00 .)(m1)(m1)
      rng (1,......;0,.11.00)(ml)(ml)
      rng (1,......;0,0 .11 .0)(m1)(m1)
      rng (1,......;0,00 .11 .)(m1)(m1)
      rng (1,......;0, .00 .11)(m1)(m1)
      rng (1,......;0,1 .00 .1)(m1)(m1)(ar,j)
      if (i .eq .j)20,10
```

* suppression of the parasite bones of the skeleton

```
20    i=j
      rng (1,......;0,1111 ..,*)(m1)(m1)(ar,j)
      if (i .eq .j)30,20
```
* at this point, m1 contains in black the skeleton

* suppression of the grains cutting the edges of the frame

```
30    nol (m2,(m2,co),un)(m2)
      rng (1,0 .....,*)(m2)(m2)
      visu (m1,m2)
      call 100
      nol (m1,(m2,co),in)(m3)
```

* at this point,
*   m1 contains all of the grains
*   m2 contains the grains cutting the edges of the frame
*   m3 contains the grains within the frame

* reset result array

```
      for n0=1,20
      n(n0)=  0
      next n0
```

* individual analysis of the grains

```
      visu (m1,m2,m3)
40    fpo (m3)(m2)
      call 100
      nol (m3,(m2,co),in)(m3)(ar,k)
```

```
      dil (m2)(m2)                    (3)
      dil (m2)(m2)
      nol (m1,m2,in)(nc,n0)
      n(n0-1)= n(n0-1)+1
      if (k .ne .0)40,50
  50  edit n                          (4)
      goto 1
```

★ subroutine of reconstruction of objects of m1 with marks in m2

```
 100  j= 0
 110  i= j
      dil (m1,m2,in)(m2)(ar,j)
      if (i .eq .j)120,110
 120  nol (m1,m2,in)(m2)
      exit
  /★
```

4.2.1 Comments

(1) When the statement HALT is encountered, the computer is put into an interactive mode. This enables the user to issue commands to the IA, such as thresholding, displaying, and writing into a memory with the joystick.

(2) The skeleton which is built here is only a subset of the median axis of the domain between the grains. It is made of points which are equidistant to at least two different grains [11].

(3) Due to the skeletonization, the domain between the grains is now only one pixel wide. Accordingly, a grain dilated by a hexagon of radius 2 pixels hits all its neighbours, and only them.

(4) Printing out the results.

ACKNOWLEDGEMENT

The author would like to express his grateful thanks to Dr J. L. Paul for his helpful comments and criticisms in an early draft of this chapter.

REFERENCES

[1] Kolomensky, E. N. and Serra, J. (1976). "La quantification en petrologie". *Bull. Assoc. Int. Geol. Ing.* **13**, 83.

] Serra, J. (1980). *Image analysis by mathematical morphology*. Academic Press, London and New York.

] Golay, M. J. E. (1969). "Hexagonal parallel pattern transformations". Trans. IEEE on Comp. **C-18**, 733–740.

] Preston Jr., K. (1971). "Feature extraction by Golay hexagonal pattern transforms". Trans. IEEE on Comp. **C-20**, 9.

] Klein, J. C. (1976). "Conception et realisation d'une unite logique pour l'analyse quantitative d'images". Thesis, Univ. de Nancy.

] Digabel, H. (1976). "Manuel d'utilisation d'AT4". Int. Rep. of the CMM, Fontainebleau

] Beucher, S. (1978). "Random processes simulation on the Texture Analyser". In *"Geometrical probablity and biological structures, Buffon 200th Anniversary"*. Lecture notes in Biomathematics, No. 23, Springer-Verlag, Berlin.

] Jeulin, D. (1979). "Morphologie mathematique et proprietes physiques des agglomeres de minerais fer du coke metallurgique". Thesis, Univ. de Paris.

•] Digabel, H. and Lantuejoul, C. (1978). "Iterative algorithms". Special issue of Pract. Metallography, No. 8, Riederer Verlag, Stuttgart.

•] Lantuejoul, C. and Beucher, S. (in prep.). "On the change of field in image analysis". (To be published in *J. Micros*).

] Lantuejoul, C. (1978). "Grain dependence test in a polycrystalline ceramic". Special issue of Pract. Metallography, No. 8, Riederer Verlag, Stuttgart.

Chapter Fourteen

GOP: A Fast and Flexible Processor for Image Analysis

Goesta H. Granlund

1. INTRODUCTION

Grey-scale and colour images with a reasonable resolution contain great amounts of information. Analysis of such images takes an excessively long time and requires large processing capabilities. For that reason fast special purpose image processors have been developed [1, 2]. Some of these processors are oriented towards use of logical operations on binary images. A common procedure is to use thresholding on an image to create a binary image where objects can be separated and described using topological transformations.

Generation of a reduced representation of an image, e.g. to a binary image, gives a large compression of the amount of information, but it also gives a great loss of information. For that reason the method can be utilized in a very limited number of situations. In fact, most situations where we would like to employ image analysis involve images with characteristics given by subtle variations in grey-scale or colour. We may have different regions described by various textures, and it is often required to detect the borders of such texture regions.

The philosophy behind the GOP image processor is to provide a very flexible and fast instrument to solve most problems in image analysis and image processing in an economical way. The processor can work on grey-scale, colour, or multi-spectral images of any size.

The processor has been designed to perform computations within the General Operator framework [3–7], in a very powerful way. However, the processor is by no means limited to this class of operations, but it can perform most arithmetical and logical operations suggested in a very efficient way. The architecture of the processor makes it in effect an MIMD (Multiple Instruction stream—Multiple Data stream) parallel machine with an extreme flexibility to be re-configured and reprogrammed for new tasks.

2. THE GOP IMAGE PROCESSOR

The GOP (General Operator Processor) image processor implements in hardware a set of particular operations described earlier [3–7], as well as most other operations suggested for image processing. The processor can be attached to a standard computer system used for image processing. A typical configuration is the one given in Fig. 1.

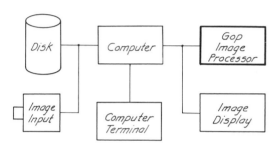

Fig. 1. Typical configuration of an image processing system using the GOP image processor.

The processor communicates with the rest of the system on the DMA channel. This normally gives a good balance between transfer speed on the DMA channel and the speed of the processor, as every picture point in general is used several times. The processing speed of the system depends upon the operations performed as well as on the characteristics of the computer system used, but it approaches 30–40 ns per input picture point for large operators. The operations that we have found to be of primary interest in image processing and analysis are of type:

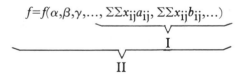

This very general form includes operations of arithmetical as well as logical type. The output is partly a function of a number of product sums which may be convolutions between mask functions and neighbourhoods of the image. The computations of these product sums is generally very time-consuming as they contain a great number of products for each sum. However, the comutation is very straightforward and needs very little flexibility.

The output is also a function of a number of parameters α,β,γ... . These parameters enable a high degree of non-linearity in the procedure. The parameters are typically functions of pictures or picture transforms which means that their value varies from one point of the image to another. The parameters can either be combined with the product sums to form any function, or they can point to different subroutines in the micro-program memory. Thus any degree of flexibility can be obtained.

The parameters can, for example, imply a variable threshold function or a dominant orientation for non-isotropic filtering.

The two types of computations described earlier require two different architectures to be performed efficiently. This gives a structure according to Fig. 2. A more detailed block diagram of the processor is given in Fig. 3.

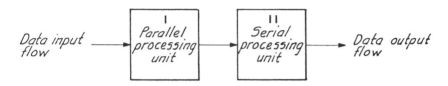

Fig. 2. Simplified block diagram of system architecture.

Part I of the processor is a reconfigurable pipelined parallel processor where data from the image segment memory and weights from the mask memory are combined in four parallel pipelines.

The image segment memory has a capacity of 32 K × 16 bits. It can be restructured to fit the current processing situation, such that up to 16 input images of any size can be involved in processing simultaneously and operators with a size up to 128 × 128 pixels can be used. Software in the external computer determine the allowable length of the image segment, which depends upon the number of images and the mask size. Usually the image segment has a length equal to the original image, and a width equal to the size of a neighbourhood. Data is moved to the processor one line at a time. There it substitutes the oldest line in a "rolling" fashion. There is a great deal of flexibility in the choice of data representation of the 16 bits word from the image segment memory, e.g. complex data, binary image data, 16-bit data.

The mask data is stored in a memory of size 16 K words of 24 bits. This storage can be restructured in a number of ways. The normal configuration is that the mask memory is divided into four sections, one section providing each pipeline with weight coefficients. In parallel with the first section is another memory of 4 K words of 15 bits. The content of this later memory points to the image segment memory selecting data points to be processed. This means that points can be picked arbitrarily within the 32 K-image segment memory to form a neighbourhood of up to 4096 points sampled in any order or arrangement. This allows, for example, masks of different sizes to be used on different

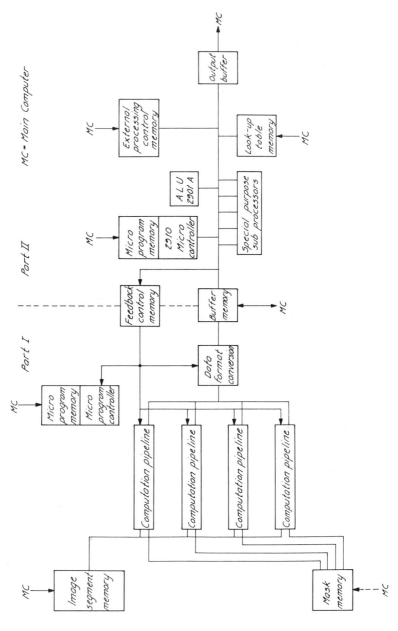

Fig. 3. Block diagram of the GOP image processor.

input image planes. The mask memory can be organized to contain up to 4096 different masks with any distribution of size within the limits of the size of the mask memory itself. These masks can be freely combined in up to 256 different mask sets. Which one of these mask sets to use can be determined by Part II, for example in response to image data intended to control the processing. Here also is a great deal of flexibility in the choice of data representation, such as complex data, 16-bit data.

In order to allow fast computation, part I of the processor uses fixed point arithmetic. However, great care has been taken not to cause errors due to overflow or underflow. Consequently, at the end of the pipeline there is a dynamic range of 48 bits.

Before data enters part II of the processor it is converted to one of three data modes: 16-bit fixed-point data, logarithmic data or floating-point data. The data mode can be selected with respect to demands on accuracy and speed in the computation in part II. The communication from part I of the processor to part II is done over a dual memory of 2 × 4 K words of 16 bits. Part I can write into one half of this memory at the same time as part II reads from the other half.

Part II of the processor has an entirely different architecture. After processing in part I, the amount of information is reduced considerably. Now a high degree of flexibility is required to combine intermediary results derived by part I. These combinations are usually highly non-linear operations, determined from one point to the other by some particular transformation of the image to process.

Part II is consequently a serial, special purpose processor. The central parts are a microprogram controller 2910 and an arithmetic logic unit 2901 B. A fast microprogram memory (access time 55 ns) of size 2 K words of 64 bits (expandable to 4 K words of 80 bits) gives a cycle time of 150 ns for the processor. All units communicate over a 16-bit bus. A special work memory of 2 K words (expandable), with possibilities of indirect addressing, facilities programming.

In order to obtain first processing of complicated algorithms, a 16 K word memory area can be used for look-up tables. A memory area of 4 K words is available for external processing control. In this memory lines from up to four images can be stored to control the processing point by point.

On the main bus are attached a number of special purpose processing units. They perform operations such as fast multiplication, scaling and shifting (up to 16 steps in one cycle), floating point operations, etc. The computation within the processor can be performed using fixed-point 16-bit representation, logarithmic 16-bit representation or floating-point representation or any desired mix of these during a particular procedure.

3. COMMON FEATURES

Part I is controlled by part II regarding what operations to perform, but does not interfere during the computations set up for a neighbourhood. However, the configuration of the pipeline can be changed after the computation, and an entirely different configuration can be set up instantaneously for a different type of computation on the same (or different) neighbourhood. This allows maximal flexibility in conjunction with high speed.

In normal processing the pipelines remain in the same mode for the whole image. Parts I and II run simultaneously at maximum speed with data exchanged over the twin buffer.

A typical operation step where maximal flexibility is needed, may go as follows:

The appropriate element from the controlling image is fetched from the memory. This data is processed in some way to determine which one of a number of pre-determined actions to take. This particular action leads to an address of the micro-controller, which goes through a procedure to activate the desired pre-set mode, to determine what set of masks to use and to give part I permission to start. Part I goes through the micro-program sequence determined by part II, thereby performing, for example, convolution with a particular mask, and returning the result to the buffer memory. Part II may now analyse this result and decide that it wants a different operation performed on a different input image (of which there may be up to 16) using a different size neighbourhood, although the computation is performed with relation to one particular output pixel. The same type of procedure is performed over again. The action decided leads to a different address in the micro-controller, which points out a different mode, a different micro-program procedure and different masks to use for part I.

The preceding procedure may sound very complicated but the computations can be performed very fast, as most possible actions can be prepared for, using look-up tables. We also have a speed of the whole procedure which is dependent upon the complexity of the procedure.

4. SOFTWARE

In order to obtain an easily workable system with a processor as flexible as the GOP, it is necessary to have a good software system. For that reason an extensive, interactive program system has been developed. The goal has been to provide program routines for most commonly occurring processing tasks, as well as to provide an attractive environment for the researcher who wants to investigate new algorithms and develop his own programs.

The program system is built around several levels of languages. The intention has again been that the program system should be easily transportable between different computers.

The highest-level language is a highly portable, interactive language, INTRAC. This language is implemented on at least five different computers, and the package is written in FORTRAN. Its purpose is to give an easy, interactive way of combining pre-coded application modules with automatic variation of parameters. This is done through the use of MACRO command files that are created and invoked by commands entered from a terminal.

The medium-level language is FORTRAN, in which the bulk of the application modules are written. The intention is that the user should be able to create his own particular application modules without too much difficulty.

The low-level language is the assembly language of the computer used, which is only utilized in very few I/O drive routines. These routines are, however, very close to the standard form for other DMA peripherals of the computer used.

As mentioned earlier, there are also assembly languages available for the creation of microprograms for parts I and II for the processor. As a rule, the average user will never have to worry about this, and he will use available modules for different modes of operation. Still, he will be able to employ the flexibility available, by specifying switches and parameters in the modules.

There are also hardware checkout program modules available. The module of main interest to the user is one that checks every function of the processor on test data, as well as the communication with the main computer. An error message is printed out in the case of error. This test program can be activated at the beginning of each work period to ensure proper operation.

Another set of available program modules can be used to create synthetic images, which is of great value in evaluation of algorithms. Programs are also available for creation and optimization of filter-functions with desired properties, and various types of mask functions.

5. MODES OF OPERATION

The GOP-processor can be run in a number of different modes, which enables different operations on images in a fast and flexible way. Filtering, edge and line detection, texture description, segmentation and labelling, relaxation, classification are some of the image processing operations the GOP-processor can do at high speed, often using the General Operator information representation.

Let us take a closer look at one example. We can run GOP in a feedback-mode for image-content dependent filtering [7]. In this case a structure like the one in Fig. 4 can be used.

Goesta H. Granlund

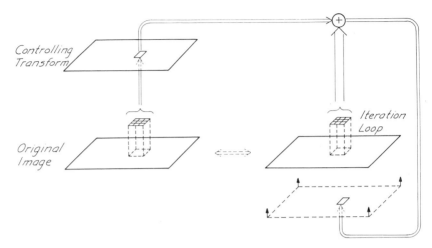

Fig. 4. Feedback structure for content dependent image filtering.

From the original image a controlling transform is computed. In the simplest case the controlling transform may be an ordinary first order transform giving the dominant orientation of structures in the image.

The original image to be filtered is now brought into an iteration loop. The image is convolved with a filter function to form a filtered image. One interesting case is when the filter function is rotationally non-isotropic, and the controlling transform determines the axis of symmetry of the filter. Figure 5 gives an example of such an iterative non-isotropic filtering of a fingerprint.

The content dependency can be developed into more complex relations with restriction rules of the same types as are used in relaxation procedures [8].

ACKNOWLEDGEMENTS

This research was supported by the National Swedish Board for Technical Development. The author also wants to express his appreciation of the enthusiastic work done by the GOP group: Dan Antonsson, Jan Arvidsson, Martin Hedlund, Hans Knutsson, Kenneth Lundgren and Bertil von Post.

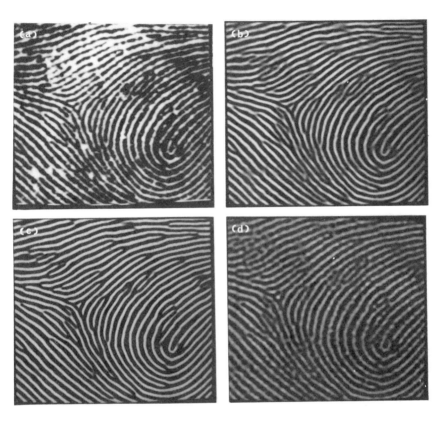

Fig. 5. Non-isotropic filtering of a fingerprint. (a) Original image; (b) result after one iteration; (c) result after two iterations; (d) a filtered fingerprint using an isotropic filter with corresponding frequency characteristics.

REFERENCES

] Kruse, B. (1973). "A parallel picture processing machine". IEEE Trans. on Comp. **C-22**, 1075–1087.

] Preston Jr., K., Duff, M. J. B., Levialdi, S., Norgren, P. E. and Toriwaki, J.-I. (1979). "Basics of cellular logic with some applications in medical image processing". Proc. IEEE **67**, No. 5, 826–856.

] Granlund, G. H. (1978). "In search of a general picture processing operator". *Comp. Graph. and Image Proc.* **2**, 154–173.

] Granlund, G. H. (1978). "On one-dimensional representation and filtering of image information". Proc. Int. Conf. on Digital Sig. Proc., Florence, Italy.

] Knutsson. H., van Post, B. and Granlund, G. H. (1980). "Optimization of arithmetic neighbourhood operations for image processing". Proc. 1st Scand. Conf. on Image Analysis, Linkoping, Sweden.

[6] Granlund, G. H. (1978). "An architecture of a picture processor using a parallel general operator". Proc. 4th IJCPR, Kyoto, Japan, pp. 1076–1081.

[7] Granlund, G. H. and Hedlund, M. (1980). "Feedback structures for image-content dependent filtering". Int. Rep. No. LiTH-ISY-I-0398, Dept. of Elect. Eng., Linkoping University, Sweden.

[8] Rosenfeld, A., Hummel, R. A. and Zucker, S. W. (1976). "Scene labelling by relaxation operations". Trans. IEEE on syst., Man. and Cyb. **SMC-6**, 420–433.

Chapter Fifteen

Evaluation of the Delft Image Processor DIP-1

F. A. Gerritsen and R. D. Monhemius

1. INTRODUCTION

In most image-acquisition, -processing and -display systems the processing part is the throughput limiting factor. Using conventional computers, image processing will be much slower than image acquisition, even in the rare cases where a small number of relatively simple operations per pixel will suffice. Throughput, however, is not the all-comprising criterion when judging the usefulness of an image-processing system. In pattern recognition and image-processing research environments the development and evaluation of new and existing methods and the application of these methods to new problems are of special interest. As routine processing of images is hardly done in such environments, throughput will mostly be regarded high enough as long as it allows efficient interactive use.

DIP-1 (Delft Digital Image Processor) is a relatively low-cost, high-performance, pipelined image processor, especially designed to be used in such research environments. Flexibility, programmability and throughput were major design considerations throughout the DIP-1 project. The processor is able to efficiently perform a large set of important basic image operations, allowing easy development of higher-level algorithms. In order to answer the demands for flexibility and programmability the basic operations were not hardwired into dedicated hardware, allowing expansion of the set of basic image operations. The machine is dynamically microprogrammable, while the pipeline consists of a number of general processing units that may be configured into an algorithm under microprogram control.

Basic image operations implemented until now include linear filtering by direct N*N convolution in the spatial domain (uniform blur, Gaussian blur, gradient or Laplace operators, inverse Gaussian filters, band filters), N*N median (and other percentiles) filtering [1], N*N edge preserving smoothing [2], histogram specification, thresholding, masking operations, erosions, dilations, contour extraction, skeletonization and runlabelling (N*N ≤ 16*16 for all neighbourhood operations). As Table 1 shows, the results of the basic image operations are visible for judgement within seconds (most within half a second), allowing concentrated human interaction.

Programming of co-ordinate transforms, fast Fourier transforms, texture analysis operators and of statistical segmentation is under way.

The design of DIP-1 started in May 1976. The initial design was presented by Gerritsen *et al.* in 1977 [3]. Working with limited manpower and budget, it took 8 man-years and approximately k$25 of parts cost to design and construct hardware, system software and firmware. DIP-1 has been operational since the beginning of 1979. The design implemented was published by Gerritsen and Aardema in 1981 [4]. This article reports on an evaluation of the DIP-1 architecture that was undertaken in view of a considered commercial redesign (DIP-2).

Table 1
DIP-1 execution times in seconds (256*256 images)

Type of operation		DIP-1 optimal pipeline	DIP-1 modular program	HP-1000 FORTRAN
Binary	3*3	0.05	(a)	23
Skeleton		0.05*W/2 (b)	(a)	75*W/2 (b)
Runlabelling		0.5	(a)	19
Pixeloperations		0.05	(a)	10
Convolution	3*3	0.18	0.8	70
Convolution	7*7	0.98	1.8	400
Convolution	15*15	4.5	5.8	700
Kuwahara filter	3*3	(a)	2	220
Kuwahara filter	7*7	(a)	7	250
Kuwahara filter	15*15	(a)	10	310
Median filter	3*3	(a)	1	120
Median filter	7*7	(a)	2	240
Median filter	15*15	(a)	4	480
Lin. relaxation	15*15	4.5 (a)	5.8	900

(a) Not implemented (yet). (b) W is the maximum diameter of the largest object.

2. HARDWARE DESCRIPTION

2.1 Overview

DIP-1 was designed as a fast and flexible processing extension to an image-acquisition, -processing and -display system described earlier (by Groen *et al.* [5] and Duin *et al.* [6]). Figure 1 illustrates the complete system. Images may be scanned by a flying spot scanner (FSS) or by television scanners (TVS, GMR-270). Images may be displayed by storing them in one of the video-speed random access image memories (DIM, GMR-270), each connected to a colour display. The image memories are also used in matching the high data rates of the TV scanners and display devices (7.2 Mbyte/s) to the lower DMA data rate of the mini-computer (0.5 Mbyte/s).

The images are processed by the HP1000 mini-computer or by DIP-1, with the HP1000 functioning as host-processor. In both cases, the user may either program in FORTRAN, combining his own algorithms with subroutines from a large image processing library (FORTRAN, HP-assembler, DIP-1), or he may try to build a high-level algorithm by applying available basic image operations using an interactive command language.

2.2 DIM image memory

DIP-1 uses an external image memory (DIM) of earlier design as working space. DIM was built with (by 1980 standards) fairly slow memory elements: 512 Intel 2102 memory chips were used (1 kbit, 1 μs cycle time—1975 NMOS technology).

The image memory is organized in an interleaved way, using eight separately controlled memory groups to allow the 7.2 Mbyte/s serial data rate needed for 50 Hz, 256*256 video input/output. The three least significant bits of the column address function as group select.

Random access cycle time (latency) as seen from DIP-1 is 1680 ns, corresponding to 6 DIP-1 micro-instructions. Using DIM in an interleaved way, block transfers between DIM and DIP-1 of rates up to 4 Mbyte/s are possible, but the microprogrammer must take care not to select a memory group for a second time within 4 micro-instructions (1120 ns). In most applications, for cosmetic reasons, transfers are limited to the periods the display does not access the image memory (which is about 50% of the time), resulting in a mean transfer rate of 2 Mbyte/s.

As DIP-1 uses DIM in a fully synchronous way, a strict timing protocol of data, addresses and control signals must be followed. Although DIP-1 was

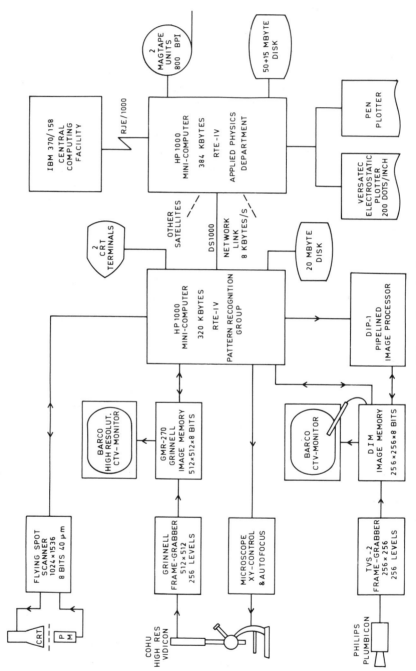

Fig. 1. The Delft Image Processing System.

originally designed to be used with two to four 256*256 image memories, only one 256*256 image memory was realized, necessitating the use of the conversion tables (see below) as an extra data-buffer inside DIP-1 for non-recursive in-place neighbourhood operations. One may also resort to using DIM as 4 128*128 image memories, trading resolution for ease of use.

2.3 DIP-1 hardware

Separate parts of DIP-1 take care of data manipulation and of DIM image memory address generation.

2.3.1 Address-loop

The address-loop (address-processor) generates the addresses in the DIM image memory of the picture elements that have to be fetched and brought to the data-loop; it also generates the addresses in the image memory of the picture elements where the results of the data manipulations have to be written.

The size of the neighbourhood, the way in which the neighbourhood scans the image and the sizes of input and output images may be varied by changing the part of the microprogram that controls the address-loop. The number of pixels that may be stored in the neighbourhood data-buffers (256 pixels each), the wordlength of the address-loop (18 bits) and the size of the image memory (in our implementation 64 kbytes) are the only practical limitations. When connected to 256 kbytes image memories, DIP-1 could handle 512*512 images. Images larger than 512*512 bring about additional paging overhead, both in hardware and in software. Apart from microprogram constants, input to the address-loop may also come from the data-loop. This way, the latter may assist in more complicated address-calculations, e.g. co-ordinate transformations.

2.3.2 Data-loop

Data is stored in the DIM image memory either as 8-bit integers or as 5-bit integers using the remaining 3 bits to code seven colours. Inside the data-loop, however, data is either handled as 12-bit mantissa, 6-bit exponent normalized floating point numbers, 12-bit integers, or as 12-bit logical fields. Conversions take place "on the fly" in the input and output modules of the data-loop. The data-loop uses hardware floating point operators, combining a large dynamic range with ease of programming and reasonably fast execution.

The data-loop consists of nine concurrently operating modules: five data manipulation modules, two 256-entry neighbourhood data-buffers, a data-input module and a data-output module (Fig. 2). The data manipulation modules include two hardware floating point alu's, a hardware floating point parallel multiplier, a (double) floating point look-up table and a look-up table for fast 3*3 binary neighbourhood processing.

These nine modules are arranged in a (reconfigurably) pipelined structure: each operand input of all of the nine modules is connected to a four-way, 18-bit wide data-selector (multiplexer, MPX—cf. Fig. 2). Each of the four data-selector inputs is connected to the output of one of the modules. More modules can choose the same output as operand. In this way, each output forms a separate bus.

2.3.2.1 Neighbourhood data-buffers

Two blocks of registers (256 locations each) supplement the data manipulation modules. These neighbourhood data-buffers (NB's) are primarily meant for storage of current neighbourhood and filter coefficients (up to 16*16) in performing neighbourhood operations, but the NB's may also be used as row or column buffers or as temporary storage for intermediate results from the data manipulation modules.

During every micro-instruction cycle, each NB may perform both a read and a write operation. The four NB read and write addresses are calculated by dedicated programmable logic. Various addressing modes are allowed: auto-increment, auto-decrement, absolute and relative addressing, but also (more complex) cyclic-buffer addressing schemes have been included, useful when updating N*N neighbourhoods by fetching N new pixels from DIM image memory.

The NB's are also directly accessible from the host computer, enabling easy exchange of microprogram input parameters (e.g. filter coefficients, scale sizes, repetition factors) or output data blocks (e.g. histograms).

2.3.2.2 Conversion tables

Two conversion tables may be used to perform transformations of integer or floating point values by table look-up in one micro-instruction cycle. The tables may be loaded by the host computer or directly from within the data-loop. Examples of their use include grey-scale transformations, thresholding, homomorphic filtering and supplying inverse values for division by multiplication.

When the conversion capabilities of the conversion tables are not needed, their ($2 \times 4 \, k \times 18$) memory area may be used to store intermediate results in full 18-bit precision, avoiding the scaling and truncation to 8-bit DIM image memory format. One may also buffer a number of input or output lines in this memory area, as required when performing non-recursive in-place neighbourhood operations (Fig. 3).

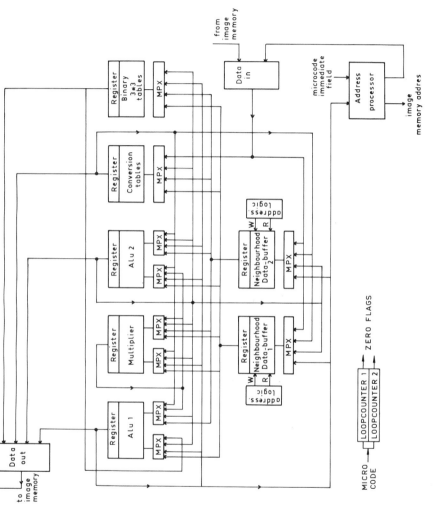

Fig. 2. Block diagram of DIP-1.

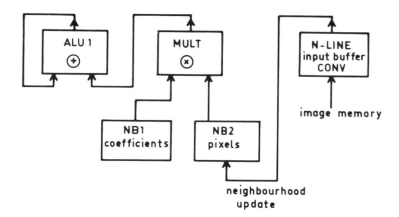

Fig. 3. Pipeline configuration for non-recursive in-place convolutions using conversion table memory space as N-line input buffer.

2.3.2.3 Binary 3*3 table module

Binary 3*3 neighbourhood processing [7] is simplified by the binary 3*3 table module of the data-loop. In two-level images, 512 possible 3*3 neighbourhood configurations exist. Using shift registers, the 9-bit binary data of every 3*3 neighbourhood is assembled and used as address for table look-up in the 512-entry, 8-bit wide table. In this way, simultaneous decisions are made (among other things) as whether the central point of the current neighbourhood is part of the 8-connected contour of an object, whether it is the endpoint of a line segment, whether it is a break pixel, changing connectivity when it is deleted, or whether it is an isolated pixel or not. The decision results may be obtained both recursively and non-recursively. Simultaneous recursive and non-recursive decisions are for instance necessary in our one-pass implementation of the topology preserving the shrinking algorithm of Hilditch [8].

A more complete description of the binary 3*3 table module has been given by Gerritsen and Aardema [4].

3. SOFTWARE SUPPORT

A wide range of programs and dedicated routines facilitate DIP-1 microprogram development, debugging and use. All support software executes on the HP1000 host computer in a multiprogramming environment. A central concept in structuring the support system was to make DIP-1 available at the FORTRAN-level. Functions such as microprogram downloading, actual parameter handling, device commands, bookkeeping and error logging are performed transparent to the user programmer. This way, the user programmer may focus upon the essence of his/her task.

The host computer functions as the primary diagnostic medium, both for microprogram debugging and processor maintenance. Maintenance programs include memory diagnostics, I/O-channel tests and utilities for memory examination. Microprogram debugging software may monitor DIP-1 operating in breakpoint mode, trace mode or in single-step/dump mode. When using the debugger software in single-step/dump mode, contents of condition registers, program status registers and output registers of all data-loop and address-loop modules are displayed on a terminal, together with the disassembled previous and current micro-instructions.

Microprogram development consists of separate stages for source-entry and -listing, assembly and linking/loading. Micro-assembler source code, consisting of 33 consecutive mnemonics for each micro-instruction, is translated by the micro-assembler program into relocatable code, using 96 bits per micro-instruction. Frequently occurring combinations of mnemonics for related fields may be specified in a "shorthand" notation. The micro-assembler also allows symbolic addressing, constant specification, comments, pseudo-instructions, and run-time specified microcode fields. Executable microcode is obtained with the linking loader. DIP-1's writeable control store is a 1 k × 96 bit RAM. In case the total amount of microcode needed exceeds this capacity, the linking loader will produce multiple control store image files that are to be swapped at run-time. Frequently needed microroutines may be made resident in the 3 k × 96 PROM part of the control store.

4. EVALUATION

By examining the microprograms written for the basic image operations we tried to

(i) identify those aspects of the hardware design that appear essential for high performance;

(ii) suggest improvements in the hard- and software design;

(iii) isolate design errors that should be avoided in a next version;

(iv) uncover costly "designer's gadgets", not significantly improving overall performance.

Before going into the details of some of our findings, we will give a summary of the most important conclusions and recommendations.

(1) The interconnection of data-loop modules by 4-way data-selectors proves to be very flexible. The interconnection graph chosen is more than adequate

for most basic image operation algorithms. The reconfigurable structure also allows fast parallel, non-pipelined processing. The 144 (= 8 × 18) interconnections between modules needed for every 2-input operator (ALU's and MULTIPLIER) and the 72 interconnections needed for every single-input operator (tables, NB's) are, however, very costly. For a cost-effective version of DIP, implementation of the interconnections by a number of high-speed time-multiplexed buses might be considered, although (dependent on the implementation chosen) this would slow down certain algorithms.

(2) The neighbourhood data-buffers are playing a central role in all micro-programs. The feature of concurrent writing and reading on independent addresses is not only useful for updating neighbourhoods while processing, but also for storage of intermediate results from ALU's and MULTIPLIER. Where mini-computers would spend most of their computing time on array-indexing, the programmable addressing logic inside the NB's relieves the data-loop ALU's of this task. NB-addressing might be made even more versatile by allowing the data-loop modules to also supply NB-addresses, supplementing the NB programmable addressing logic to enable data-dependent addressing. The "smart" jumps emulating a cyclic-buffer are very useful in updating N*N neighbourhoods by supplying N new pixels. A next version should use data-buffers as large as 8 kwords, to allow storage of 16 lines of 512 pixels (see also recommendation 7).

(3) By using the two hardware loop counters (Fig. 2), very fast loops may be programmed easily, as the ALU's are relieved of the loop counting task. Some algorithms need 3 or 4 loop counters, while only two hardware loop counters have been provided. This need for more loop counters might be met by allowing the loop counters to be saved and restored by the data-loop modules, rather than providing 4 loop counters.

(4) Most algorithms do not use data-dependent addressing of the image memory. For these algorithms, the address-loop might be replaced by a number of accumulators. The general purpose registers of the address-loop should then be included in the data-loop ALU's. This might be easily realized by using bit-sliced processing elements for the ALU's, instead of MSI logic.

(5) The 12*12 bit parallel multiplier used in the MULTIPLIER unit might be replaced by a 16*16 bit multiplier-accumulator chip.

(6) The microprogram development software (and built-in monitoring hardware) facilitates the task of the microprogrammer substantially.

(7) The access to the external DIM image memory is not satisfactory. In a next version, either the image memory should have a better cycle time, or the neighbourhood data-buffers should be chosen large enough to hold N lines (see also recommendation 2). This point will be discussed extensively below. Furthermore, a next version should use at least two, preferably four 512*512 image memories.

(8) The choice for the 18-bit floating point data-representation should be reconsidered in the light of the fact that 512*512 images (and even larger) are used more and more, mostly necessitating the use of correspondingly larger convolution filter neighbourhoods. This requires a better precision. A 24 to 32-bit integer representation would probably be accurate enough for most operations. This point will be discussed in more detail below.

4.1 DIM image memory access

The present architecture allows optimal pipelining of neighbourhood operations in the sense that image memory access and data manipulations may be performed in a continuously working pipeline (Fig. 4, Fig. 5a).

It is, however, rather difficult to write microprograms that perform input from image memory and neighbourhood processing in a concurrent, pipelined way, while it is virtually impossible to write such microprograms that will perform optimally for varying neighbourhood size. As the (rather slow) image memory demands a strict timing of data, addresses and control signals, the timing of the data processing will mostly have to be adapted to the image memory timing requirements. Choosing a different neighbourhood size will

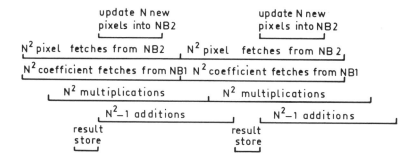

Fig. 4. Continuously working pipeline for N*N convolutions using a N*N neighbourhood data-buffer (NB). Disadvantage: modular programming not possible for optimal data-rate.

change the address-stream to the image memory, demanding readjustment of the timing to retain the optimal data rate. In other words: for almost every neighbourhood size a new, optimized microprogram has to be written.

In practice, microprogrammers refuse to pay this price for maximum speed. Instead, input is not performed concurrently with neighbourhood processing, but input, pipelined neighbourhood processing and output are alternated by using subroutines for image memory access. These I/O-routines also take care of the buffering of input lines necessary because of the absence of a second DIM image memory for storing results of computations.

N*N neighbourhood operations (variable N) are programmed by just supplying a micro-subroutine performing the operations to be repeated for every neighbourhood. All necessary scanning of the full size image is then taken care of by the I/O-routine. Neighbourhoods are updated (in a transparent way) from a N-line input buffer in the conversion table memory space into one of the neighbourhood data-buffers, the subroutine specifying the neighbourhood operation is entered and after completion of the processing of the neighbourhood, the result is written to a 1-line output buffer in the conversion table memory space. The updating of the N*N neighbourhood in the NB requires only N pixels to be copied from the input buffer to the NB, as the NB may be addressed in a cyclic buffer mode.

Memory fetches beyond the edges of the image are also handled by the I/O-routine: "mirrored" data is supplied, emulating a symmetrically periodic image.

Obviously, use of these I/O-routines substantially improves microprogramming productivity, but unfortunately also results in some degradation of machine performance, as input, processing and output are alternated instead of performed concurrently. Pipelined neighbourhood processing will mostly take N*N + S micro-instructions per N*N neighbourhood (S being the pipeline start-up time). Using the I/O-routines, only one DIM image memory input cycle and one DIM output cycle are needed per N*N neighbourhood, plus 3N instructions to update the neighbourhood from the input buffer to NB. This overhead is small for N*N larger than (say) 7*7, but certainly not for N smaller than 5. The conversion tables were not designed to function as input/output buffers. The resulting inelegant addressing of the tables for this purpose needs three instructions (and not just one) per transfer from conversion tables to NB.

In a next version of DIP, the size of the neighbourhood data-buffers should be chosen large enough to let the NB's function as N-line input buffer. In this way, the 3N overhead would be completely eliminated, resulting in an attractive combination of almost optimal pipelining (for neighbourhood operations) and modular programming—in spite of the slow image memory (Fig. 5; and also see recommendations 2 and 4).

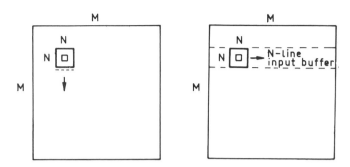

Fig. 5. (left) When using a N*N neighbourhood data-buffer, N fetches and a store are needed per neighbourhood. Two image memories needed. (Neighbourhood moving down to allow fast row fetches.) (right) When using a N-line input buffer, the number of memory accesses per neighbourhood is reduced to two: a fetch and a store. In-place operations possible. Combination of optimal and modular programming.

4.2 Accuracy

DIP-1 uses hardware floating point operators (12-bit mantissa, 6-bit exponent). The dynamic range of the floating point numbers is (roughly) from $\pm 1 \times 10^{-10}$ to $\pm 2 \times 10^{9}$. Every multiplication or addition is performed in one micro-instruction cycle (280 ns).

The most important advantages of integer representation are the high speed attainable and the relatively simple hardware. An important disadvantage is the scaling problem the programmer needs to solve. This is the strong point of the floating point representation: using the same number of bits (the same wordlength) the dynamic range will be much larger than when using the integer representation. The scaling problems, however, are traded for accuracy problems. Another disadvantage of the floating point representation is the fact that the processing latency is higher because of the necessary shifting for predenormalization and postnormalization.

Originally, DIP-1 was intended to handle neighbourhoods with sizes up to 8*8. It was at that time the decision was made that the floating point representation should be 18 bits wide with a 12-bit mantissa. It was only in a much later stage of design that 16*16 neighbourhoods were decided upon.

We felt the mantissa width might be somewhat limited for performing 16*16 convolutions of 8 bits/pixel images, which led us to compute the theoretical worst case error in convolutions using DIP-1. We also experimentally compared convolution results of the HP1000 32-bit (24 + 8) floating point format with DIP-1's results for a number of filters.

From the theroretical worst case errors and from the experimental comparisons it may be concluded that when using neighbourhoods larger than 8*8 and filters with a very large dynamic range, one should be cautious in interpreting DIP-1 convolution results as the cumulation of errors may be considerable.

Apart from the fact that errors are amplified by the computations (!) the following sources of error may be distinguished:

(1) Conversion of filter coefficients or other constants from 32-bit HP1000 format to 18-bit DIP-1 format. The HP1000 is functioning as DIP-1's host. The host performs functions as down-loading DIP-1's microprograms and the filling of DIP-1's memories with constants and parameters. It will be clear that the floating point numbers computed by the host will have to be converted to the smaller DIP-1 format. In this conversion a number of significant bits may be lost.

(2) Post-multiplication truncation errors.

(3) Denormalization of operands before addition/subtraction.

(4) Post-addition/subtraction truncation errors.

(5) Conversion of internal floating point format into image memory 8-bit integer format (all negative values are mapped to 0, all values larger than 255.2^{SCALE} are mapped to 255).

A number of modifications have been proposed which would halve the total error:

(1) Using round-off instead of truncation in converting from HP1000 FP-format to DIP-1 format. (As this conversion was done in HP-assembler software this modification was easily realized and indeed showed a minor improvement.)

(2) Using round-off instead of truncation in ALU's, multiplier and data-output module. The most simple implementation seems to be a ROM-rounding.

(3) Round-off of the denormalized operand before addition or subtraction instead of truncation.

For DIP-2 however, a 24 to 32-bit integer format, a 16*16-bit multiplier-accumulator and the use of block-scaling are considered, which will allow higher processing speed and more accuracy. (See also recommendations 4 and 5.)

5. CONCLUSIONS

DIP-1 is a relatively low-cost, high-performance image-processor, especially suited for interactive use in pattern recognition and image processing research environments, where DIP-1's flexibility, programmability and ease of use are certainly as important as its speed. Although image memory access and accuracy could be improved in a redesigned version, the extensive (and successful) use of DIP-1 over the past two years has validated the main features of its architecture.

ACKNOWLEDGEMENTS

The authors wish to express their appreciation to prof. dr. ir. C. J. D. M. Verhagen, prof. ir. G. L. Reijns, prof. ir. C. H. Eversdijk, drs. W. G. van den Berg, dr. ir. F. C. A. Groen, dr. P. W. Verbeek, dr. ir. R. P. W. Duin, dr. I. T. Young, ir. L. G. Aardema, ir. H. Sennema, ir. J. Spierenburg, R. J. Ekkers and J. van den Hooven for their numerous contributions and suggestions and for their stimulating discussions.

REFERENCES

1] Huang, T. S., Yang, G. J. and Tang, G. Y. (1978). "A fast two-dimensional median filtering algorithm". Proc. Conf. on Patt. Recog. and Image Proc., Chicago, pp. 128–131.

2] Kuwahara, M., Hachimura, K., Eiho, S. and Kinoshita, M. (1976). "Digital processing of RI-angio-cardiographic images". In *Digital processing of biomedical images*, Preston Jr., K. and Onoe, M. (eds). Plenum Press, New York, pp. 187–202.

3] Gerritsen, F. A., Spierenburg, J. and Sennema, H. (1977). "A fast, multipurpose, microprogrammable image processor". Proc. SITEL-ULG Seminar on Patt. Recog., Liege, Belgium, pp. 611–618.

4] Gerritsen, F. A. and Aardema, L. G. (1981). "Design and use of DIP-1: a fast, flexible and dynamically microprogrammable pipelined image processor". To be published in Patt. Recog. **13**.

5] Groen, F. C. A., Verbeek, P. W., Zimmerman, N. J., Heusdens, J. J., Snidjer, R., Vis, P. A., Wichern, P. H. M., Wolff, H. J. G., Hoeksma, Th., Knoop, W. L. and Beekum, W. T. van. (1976). "A universal measurement and processing system for two-dimensional images". Proc. Imeko VII Cong., London, Acta Imeko, pp. 259–267.

5] Duin, R. P. W., Gerritsen, F. A., Groen, F. C. A., Verbeek, P. W. and Verhagen, C. J. D. M. (1980). "The Delft image processing system, design and use". Proc. 5th ICPR, Miami Beach, USA, pp. 768–774.

7] Preston Jr., K., Duff, M. J. B., Levialdi, S., Norgren, P. E. and Toriwaki, J.-I. (1979). "Basics of cellular logic with some applications in medical image processing". Proc. IEEE **67**, No. 5, pp. 826–856.

8] Hilditch, C. J. (1969). "Linear skeletons from square cupboards". In *Machine intelligence 4*, Michie, D. and Meltzer, B. (eds). University Press, Edinburgh, pp. 403–420.

Chapter Sixteen

Système Multiprocesseur Adapté au Traitement d'Images

J. L. Basille, S. Castan and J. Y. Latil

1. INTRODUCTION

SY.MP.A.T.I. has been conceived in order to satisfy the necessities of the different researches in Pattern Recognition run in our laboratory such as: remote sensing, neph analysis, biomedical applications, pictorial data compression and transmission and so on. We wanted a system which could:

(a) input/output Picture data at TV rates;
(b) have image manipulations facilities;
(c) process complete applications.

In order to meet these requirements we have defined a two level parallel system, built around one or several memories. The first level is of a Multiple Single Instruction stream Multiple Data stream type. It makes it possible to process plain treatments directly at memory level. The second level concerns all the system and is of a Multiple Instruction stream Multiple Data stream type. Standard processors connected to the image memories make it possible to process more elaborate treatments.

SY.MP.A.T.I. can be connected to a host computer, for example the Telemecanique T1600 mini-computer on which we are working, particularly with an interactive software adapted to image processing [1], and on which we are developing a PASCAL-like language with image processing specifications [2].

2. GENERAL ARCHITECTURE

The processors used for treatment are all standard in order that a procedure may be processed on any free processor. They take advantage of the easy and fast access to information due to the structure of the image memories (see below) they are connected to. They may receive (resp. send) information from (resp. to) each other either by using an inter-processor bus or by a communication loop. They may also receive (resp. send) information from (resp. to) the image memories by using a fast bus.

Specialized modules allow high speed exchanges with external devices. The whole system is managed by a resource allocator (Fig. 1).

2.1 The resource allocator

This takes advantage of the same communication facilities as the standard processors. On a procedure request, it may run or stop another procedure on any free processor by using the interruption bus. It also manages the specialized modules, the transfer module and the memories by using the command bus.

2.2 The different communications

2.2.1 The fast bus

The specialized modules and the image memories are connected to the fast bus and so may exchange information at 10 MHz. This makes it possible to get an information window of any shape out of an image (eventually the whole image) in less than 20 ms. It also gives the possibility to analyze an image with a TV camera or to synthetize an image on a TV monitor in real time.

2.2.2 The inter-processor bus

The local memories of the standard processors and the memory of the resource allocator are connected to the inter-processor bus driven by the transfer module. Therefore it permits the following possibilities:

(a) sending the operating code of a procedure to a free processor before running it;

(b) exchanging information blocks between two processors. For instance the results of one procedure called by another;

(c) exchanging information between a processor and an image memory as the inter-processor bus is connected to the fast bus.

Such exchanges are made transparent for the processors concerned as they are stopped by the bus manager in order to get the maximum input/output rate.

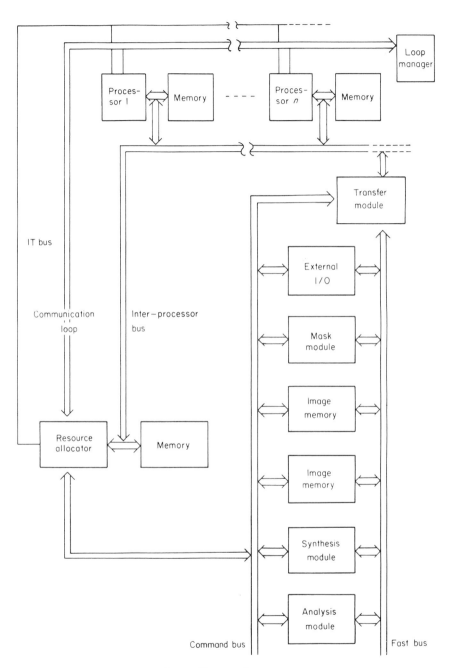

Fig. 1. The general structure.

2.2.3 The communication loop

This is used when a processor has to send only a short piece of information to another one. It saves the resource allocator requesting to send short parameters or to receive intermediate results.

2.3 Specialized modules

In addition to the image memories and the binary memories, SY.MP.A.T.I. will have three modules for image input–output:
  (a) An image analyser to digitize an image from a TV signal.
  (b) An image synthetizer to visualize image contents on a TV monitor.
  (c) An input–output module to manage the connection with another computer, for instance.

3. MEMORY STRUCTURE

The logical and physical structure of the SY.MP.A.T.I. system specific memory is constructed on the model given by the logical organization of the images. Information access is made easier by the use of "window notion". A processing unit on each of the 16 blocks constituting a memory makes it possible to compute, in an associative way, some micro-programmed procedures such as local operators processed on part or the whole of the image.

3.1 Architecture

First, two 512 × 512 byte memories will be used in SY.MP.A.T.I. to store pictorial information.

3.1.1 The memory

Each memory consists of sixteen 16 K byte blocks such that the block n contains all the n (modulo 16) columns of the image. The blocks are connected to a fast shifting loop, so as to transmit information from one block to another (Fig. 2).

This shifting loop is also used to input or output information at a fast rate compatible with a TV frequency signal. It gives an average memory access time less than 100 ms although the dynamic R.A.M. we chose to realize a memory is rather slow (1 μs).

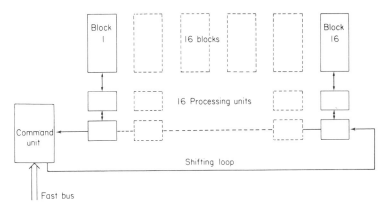

Fig. 2. An image memory.

A memory may be accessed either:
(a) by image scanning (i.e. sequential scanning of the successive lines);
(b) by TV scanning (i.e. sequential scanning of the odd lines alternating with sequential scanning of the even lines).

3.1.2 The memory processing units

The memory structure with sixteen 16 K byte blocks connected to a fast shifting loop is particularly well adapted to the neighbourhood access. Effectively the block n contains all the n (modulo 16) columns of the image, so:
(a) the close column-neighbours of (I, J) are in the same block;
(b) the close-line-neighbours of (I, J) are in the neighbouring $(J - 1)$ (modulo 16) and $(J + 1)$ (modulo 16) blocks of the J (modulo 16) block containing (I, J) (see Fig. 3).

That is the reason why we have included a processing unit on each of the 16 blocks. These 16 processing units work in an associative way and so may process local operators in a fast way.

Each processing unit is composed of an arithmetic and logical unit, an 8 register scratch Pad, and an indicator set (Fig. 4).

3.1.3 The command unit

The whole of the 16 blocks of a memory and of the connection to the shifting loop is managed by a command unit made of 8 automata. The procedures computable on a memory are microprogrammed with two level micro-instructions:
(a) the short micro-instructions give the sequence of the micro-program;

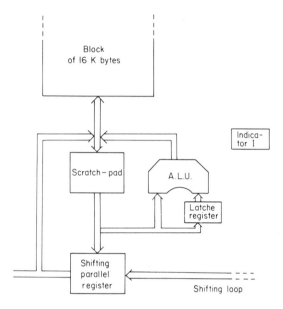

Fig. 3. A processing unit.

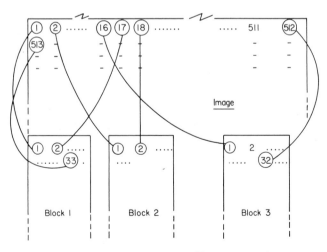

Fig. 4. Image location/memory address correspondence.

(b) the long micro-instructions command the different parts of the memory: memory access, ALU, shifting loop, command unit, and sequence the short micro-instructions for control breaks.

3.2 Information access

3.2.1 The windows

Two types of window are defined. The first is the simplest. It is a rectangular sub-matrix of the matrix of the considered image pixels. So it is sufficient to provide the co-ordinates delimiting this rectangle in the original window which may be the whole image. The second gives the possibility of defining regions of any shape. In this case, in order to keep the line-by-line image organization, it is necessary to have a rectangular window surrounding the considered region.

3.2.2. The masks

A window is realized with binary mask for the first type, or with two combined binary masks (as shown Table 1) for the second one.

Table 1

First mask	Second mask	Transfer
0	0	Masked
0	1	Set to 0 if reading
1	0	Masked if writing
1	1	Transferred

Region labelling of an image provides a good example of the use of the window. Consider one image memory with the pixel grey-values of an image and another one with, for each pixel, the number of the region it belongs to. It is then possible to get the grey values of all the pixels of any region in (only) one 20-ms scanning.

A mask may be either:

(a) one of two 512×512 bit memories previously charged;

(b) an image memory. In this case the image content is not modified as only the result of its binarization is used.

Thus we have three input–output possibilities:
(a) a whole 512 × 512 byte image;
(b) a rectangular window by using one mask scanned in parallel with the image memory containing the information;
(c) a window of any shape by using two masks scanned in parallel with the considered memory (see the above transfer table).

3.3 Expected performances

Some classical algorithms have been evaluated for a whole 512 × 512 pixel image with 256 grey levels. The results are as follows:

gradient with 4 neighbours	140 ms
gradient with 8 neighbours	200 ms
correlation with a 5 × 5 binary mask containing 10 "1"	100 ms
image transposition	300 ms

In order to provide a computation speed comparable with other parallel structures, the computation time per pixel seems to be a good unit. Then we can see that the above algorithms give us a time of about 1 μs/pixel or even less, with only 16 processors implemented.

4. CONCLUSIONS

Our experience in Pattern Recognition especially in Image Analysis has led us to conceive this double structure. The image memory organization allows an easy access to information, thanks to the window notion. Furthermore the adjunction of processing units running in an associative way gives the possibility of computing some procedures at memory level in a very short time. On the other hand, the use of standard processors connected to these memories makes it possible to increase the processing speed of more complex treatments, through the splitting into procedures that can be run in parallel and benefit from the facilities of access to information stored in the memories.

REFERENCES

[1] Argilas, F. C. and Castan, S. (1979). "Système interactif de traitement d'images (S.I.R.I.U.S.)". Colloq. Nat. sur le traitement du signal et ses applications, Nice.
[2] Argilas, F. C., Castan, S. and Gleizes, A. M. (1979). "Un langage spécialisé dans le traitement d'images". Congrès AFCET/IRIA Reconnaissance des formes et intelligence artificielle, Toulouse.

3] Bovelt, J. P. and Baer, J. L. (1968). "Compilation of arithmetic expressions for parallel computation". IFIPS congress, 1968.

4] Barnes, G. H. *et al*, (1968). "The ILLIAC IV computer". Trans. IEEE on Comp. **C-17**, 746.

5] Basille, J. L. (1976). "Traitement d'images numériques. Application: système interactif de caryotype". Thesis, Université de Toulouse.

6] Basille, J. L. and Castan, S. (1977). "Systèmes interactifs de caryotypie". Colloq. Nat. sur le traitement du signal et ses applications. Nice.

7] Basille, J. L. and Castan, S. (1976). "A fast karyotyping approach". NATO Advanced Study Institute on digital image processing and analysis, Bonas.

8] Basille, J. L., Castan, S. and Latil, J. Y. (1979). "Structure logique et physique de l'information dans un multiprocesseur adapté au traitement d'images". GRETSI 7eme colloque sur le traitement du signal et ses applications. Nice.

9] Bernstein, A. J. (1966). "Analysis of programs for parallel processing". Trans. IEEE on Elect. Comp. **EC-15**, 757–763.

0] Comte, D. and Durrieu, G. (1975). "Techniques et exploitations de l'assignation unique, **5-7-75**. Contrat SESORI 74.167.

1] Conway, M. E. (1963). "A multiprocessor system design". Proc. AFIPS Fall Joint Comp. Conf., 1963.

2] Cordella, L., Duff, M. J. B. and Levialdi, S. (1967). "Comparing sequential and parallel processing of pictures". Proc. 3rd IJCPR, Coronado, USA, pp. 703–707.

3] Duff, M. J. B., Watson, D. M., and Deutsch, E. S. (1974). "A parallel computer for array processing". Proc. IFIP Congress 1974, Stockholm, Sweden, pp. 94–97.

4] Duff, M. J. B. (1978). "Parallel processing techniques". In *Pattern recognition. Ideas in practise*, Batchelor, B. G. (ed.). Plenum Publishing Co., New York.

5] Godsen, J. (1966). "Explicit parallel processing description and control in programs for multi- and uni-processor computers". Proc. AFIPS Fall Joint Comp. Conf. **29**, 651–660.

6] Holland, J. H. (1959). "A universal computer capable of executing an arbitrary number of sub-programs simultaneously". Proc. Eastern Joint. Comp. Conf., 1959.

7] Kruse, B. (1976). "The PICAP picture processing laboratory". Proc. 3rd IJCPR, Coronado, USA, pp. 875–881.

8] Lamport, L. (1974). "The parallel execution of DO loops". *Comm. ACM*, **17**, 83–93.

9] Levialdi, S., Isoldi, M. and Uccella, G. (1980). "Programming in PIXAL". Proc. IEEE Workshop on Picture Data Description and Management, Asilomar, USA, pp. 74–79.

0] Levialdi, S., Maggiolo-Schettini, A., Napoli, M. and Uccella, G., (1979). "Considerations on parallel machines and their languages". Workshop on High-level Languages for Image Processing, Windsor, U.K.

1] Proc. Sagamore Comp. Conf. on Parallel Processing, Syracuse, 1973.

2] Stokes, R. A. (1967). "ILLIAC IV: route to parallel computers". *Elect. Design* **26**, 64–69.

3] Swan, R. J., Fuller, S. H. and Siewiprek, D. P. (1977). "A modular multi-processor". Proc. AFIPS Comp. Conf. **46**.

4] Timsit, C. and Boudarel, R. (1977). "PROPAL II: une nouvelle architecture de calculateur adaptée au traitement du signal". Colloq. Nat. sur le traitement du signal et ses applications, Nice.

Chapter Seventeen

The EMMA System: An Industrial Experience on a Multiprocessor

R. Manara and L. Stringa

1. INTRODUCTION

ELSAG of Genova, Italy, started working in the area of multicomputer systems in 1972, in connection with pattern recognition problems applied to automatic mail sorting.

The approach used, however, was very general, the basic idea being that of designing not a special purpose computer but rather a system that could be used in different areas requiring high real time performances such as image processing, speech recognition, artificial intelligence and so on.

The first prototype of the system, called EMMA®,★ became operational in 1975 [14, 15]. The name EMMA is an acronym for Elaboratore Multi Mini Associativo (Associative Multi Mini Processor). Its first application was made in SARI (Sistema Automatico Riconoscimento Indirizzi), an optical address reader developed for the Italian Postal Administration and subsequently adopted also in France while being evaluated also by other Postal Administrations, including the US Postal Service (in March 1980 a SARI was installed in Philadelphia and successfully passed field tests in September 1980).

Other applications so far developed or under development concern the reading of microfilm rolls (where each picture is the image of a bank of telephone counting meters), the reading of bubble chamber photographs and automatic speech recognition.

The architecture of EMMA is of the "hierarchical cluster type" (a fourth type added to the three categories in Enslow's classification [5]), thus featuring

---

★ EMMA is a registered trade name by ELSAG, Genova (Italy). Its implementation is covered by US patent Nr. 4099233.

a hierarchical structure composed of various communication channels, each identified by a family (cluster) of processors (each with its own memory) and I/O units.

Systems now operating are made up from three to four families for a total of 60–70 processors, but larger systems, made up from six families for a total of 120 processors, are planned for 1981.

The following will be devoted to EMMA's architecture, to the main aspects of its operating system and finally to some of its applications.

2. EMMA'S SYSTEM ARCHITECTURE

Multicomputer systems are typically of the SIMD type (**S**ingle **I**nstruction stream, **M**ultiple **D**ata stream [see 6]) that are strongly oriented towards the solution of specific problems [see for instance 20] and of the MIMD type (**M**ultiple **I**nstruction stream, **M**ultiple **D**ata stream [see 6]) that are suited for more general purpose applications [7, 9, 19, 21].

EMMA belongs in this second category. It is based on a structured set of processors that may work independently from one another. Its special purpose orientation derives more from the structure of the operating system than from the arrangement of hardware. Actually EMMA is a highly modular and easily reconfigurable system, and is therefore suited to a large range of applications.

The basic element of EMMA is the **A**ssociative **U**nit (AU), consisting of a 16-bit minicomputer whose CPU has been specifically designed and built to execute at very high speed the usual arithmetical and logic operations. (Standard execution times range typically from 0.6 to 1 μs.) The CPU of an AU, besides the standard features found in a minicomputer, has some special instructions (and the corresponding hardware) to compare strings of bits and to derive parameters such as Hamming's distance. Each AU has an associated memory module that may range in size from 4 K to 32 K words and is connected to the CPU and to a *common bus* that connects together the AU's. The resulting structure is sketched in Fig. 1.

The common bus is one of the most interesting characteristics of EMMA. It works basically in a DMA mode and was optimized for block transfers thus reaching the speed of 1 M word/s. The bus is controlled by the Data Exchange Co-ordinator (DEC). Processors normally use their local memory. If a processor wants to exchange information with another processor, it sends a request to DEC. As soon as all requests of higher priority have been satisfied, DEC obtains from the requesting AU the parameters concerning the exchange (such as name of sender, name of receiver, pointers to data block, length of block) and co-ordinates a DMA transmission. When the transmission is over, DEC signals all processors involved (it should be noted that the processor

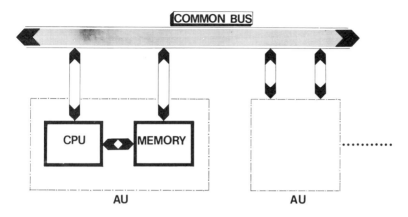

Fig. 1. Architecture of EMMA.

requesting the exchange of information does not necessarily coincide with either the sender or the receiver).

A set of AU's connected through a bus forms a *family*, which may have up to 128 AU's. All systems having a common bus structure show transmission saturation values that depend on the percentage of accesses to non-local memories and on the amount of messages exchange on the common bus [1, 9], so a family of AU's may originate a heavy load for EMMA's common bus. Therefore if an application involves both massive computation and intensive exchange of information, the system may be structured as a set of families, interconnected but using independent buses (Fig. 2).

These multibus structures are controlled by a MONITOR, which in most cases is a standard minicomputer used to interface the system with standard peripheral units and to implement fault diagnosis procedures. An EMMA system is now under development, using as MONITOR one of the AU's instead of the standard minicomputer.

There is no upper theoretical limit to the number of families, so that the size of the system may be optimized for the application, assuming it shows a convenient degree of parallelism. Systems now operating are made up from three to four families for a total of 60–70 processors. EMMA's hierarchical structure is shown in Fig. 3. The figure shows that each family may be divided into subfamilies, each formed by the section of the family bus physically corresponding to one rack. Special peripheral units, typically instrumentation, may be interfaced by means of the I/O channels of the AU's, or directly through the bus (Fig. 4).

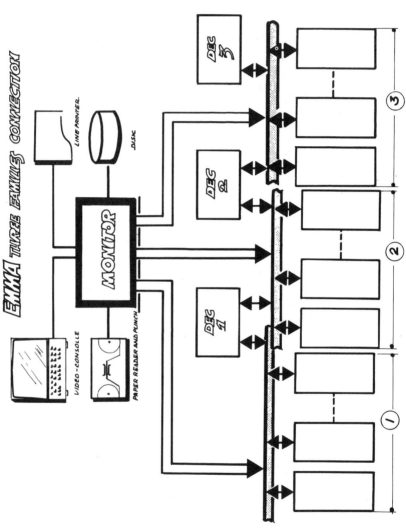

Fig. 2. Functional organization of the EMMA system.

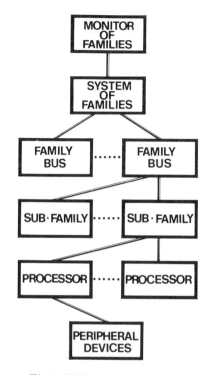

Fig. 3. EMMA hierarchial structure.

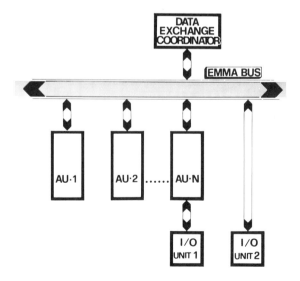

Fig. 4. I/O interfacing.

3. REMARKS ON EMMA'S OPERATING SYSTEM

3.1 General structure

EMMA's operating system is structured into levels that mostly reflect the hierarchical structure of hardware. If each level is associated with an "extended machine" [11] the resulting structure is illustrated in Fig. 5.

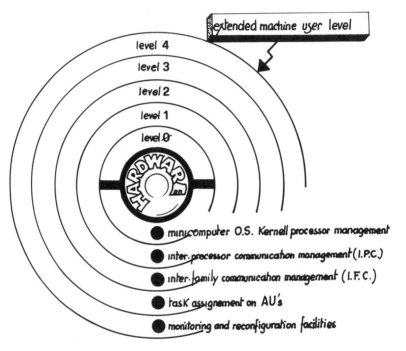

level 4
level 3
level 2
level 1
level 0

extended machine user level

● minicomputer O.S. Kernell processor management

● inter-processor communication management (I.P.C.)

● inter-family communication management (I.F.C.)

● task assignment on AU's

● monitoring and reconfiguration facilities

Fig. 5. Structure of software.

Level 0 implements some basic functions that do not require communications among the AU's, such as management of interrupts and supervisor calls. At this level the system is seen as a collection of independent units.

Levels 1 and 2 manage communications respectively within a family and among families. They supply synchronization primitives and allow communications to take place with the minicomputer used as monitor. These two levels form most of the core of a distributed operating system residing in each AU. Level 3 resides in the monitor and is responsible for the activation and distribution of processes among the AU's, taking into account the indications of the application designer and the resource requirements of the various tasks.

This level saves all the information necessary for reconfigurations. Level 4 has modules residing in the monitor and modules distributed among the AU's. They are used for fault-diagnosis and for system reconfiguration [18].

We will now examine in more detail the main functions of the operating system.

3.2 Processes: definition, synchronization, mutual exclusion

One of the advantages most frequently mentioned for multiprocessor systems is the availability of larger hardware resources that makes acceptable a less optimized use of them, which in turn simplifies the control software. Given the decreasing cost of hardware and the increasing cost of software the selection of a multiprocessor system may then turn out to be cost effective. As a consequence the conceptual difference between processor and process may disappear: if processors are abundant and cheap, the set of abstraction (processes) may coincide with the set of physical objects (processors).

According to this philosophy, each of EMMA's AU's has only one associated process, which consists of the execution of a procedure called COROUTINE. A COROUTINE consists of a set of modules having the structure outlined in Fig. 6.

Mixing processes and processors simplifies also the system's primitives; for instance, only one synchronization structure is required instead of the two (lock and semaphore) found in other systems [2, 4, 8].

Furthermore as many problems are solved by procedures which may be decomposed into blocks, executed in a sequence and weakly connected from the point of view of data structures, it is possible to view the application software structured as indicated in Fig. 7 (which, however, is a simple example that may be easily generalized).

If one module is associated to each block, A and B, and two processors are available, the system may be configured as a "pipeline" so as to exploit the available computing power. The resulting behaviour may be modelled by the Petri net [12] of Fig. 8.

Two remarks are in order when dealing with the use of synchronization primitives to solve mutual exclusion problems. First of all, EMMA's design philosophy allows any processor to exchange data with other processors, restricting, however, its addressing capability to its own local memory. If a process requires more memory than that available on its AU, it is broken into several processes assigned to different AU's, that synchronize each other by means of messages. Therefore there are no common data structures: a process that wants to operate on a non-local data structure must ask another process to do it.

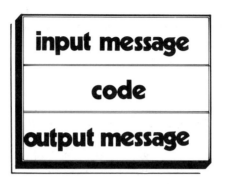

Fig. 6. Module organization.

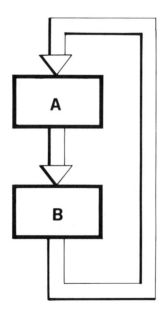

Fig. 7. Structure of application software.

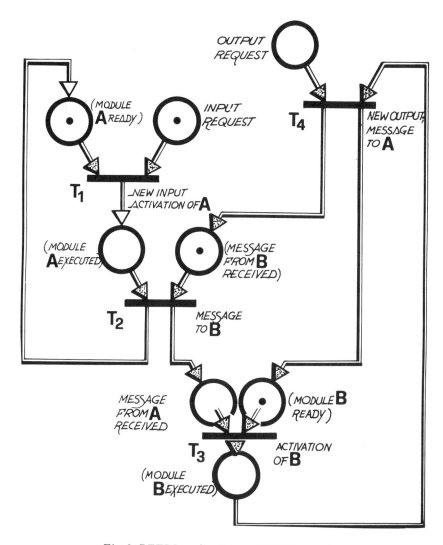

Fig. 8. PETRI net description of EMMA operation.

This approach is very similar to Brinch Hansen's model for distributed processes [2]. Mutual exclusion problems arise therefore only in connection with the handling of message buffers, and are solved directly by the bus controller since it has, during transmission of a block of words, exclusive control over the bus and the memories of transmitting and receiving AU's.

4. EMMA'S APPLICATIONS

4.1 Applications already operating

4.1.1 SARI*

Developed for the Italian Postal Administration, SARI [16, 17] can process over 50,000 mail items per hour, recognizing typed or hand printed addresses (ZIP code, town name, district name or initials). A prototype of SARI was in operation for 12 months since July 1976 in an Italian Post Office and processed several millions of mail items in a series of exhaustive tests. SARI systems are presently operational in several primary centres of the Italian Postal Administration; each system is an EMMA with 45 AU's, subdivided into 3 families.

The process of address recognition may be outlined as follows. The letter, coming from a feeder, is scanned by a reading head and after an A/D conversion, is analysed by a set of processors using software modules that perform noise reduction, line searching, line selection, word segmentation, type searching and type recognition. A set of possible outcomes is built and then matched with the proper address repertoire: searches are very fast on account of the special associative hardware.

4.1.2 SARI—France

In this application, derived from section 4.1.1 above and developed for the French Postal Administration, an EMMA system is used with 65 AU's, subdivided into 4 families.

Two sets of programs can be loaded and executed on a system: the first set is analogous to that of SARI for the Italian Postal Administration, and performs address recognition in the address line containing town name and ZIP code (the bottom line); the second set performs also the recognition of city district and street names in the last-but-one address line.

The system became operational in September 1977 and an extensive testing has been successfully performed at the French Postal Administration Centre in Arcueil (Paris). Now a SARI is operative in Bordeaux and six other systems have been ordered by the French Postal Administration.

A SARI system was also installed in the United States Postal Office of Philadelphia and successfully passed field tests in September 1980.

---

* Sistema Automatico Riconoscimento Indirizzi (Automatic Address Recognition System).

4.1.3 MAORS*

MAORS is a system performing two functions:
 (i) Reading of microfilm rolls each photogram of which is the image of a telephone counter panel (10×10 grid) at the speed of 200,000 photograms/hour.
 (ii) Reading of two lines of numeric characters in credit certificates, at the speed of 60,000 certificates/hour.
For the two functions described above an EMMA system is used with 50 AU's subdivided into 3 families.

Photogram quality is rather poor and the isolation of the useful information is very difficult since one has to deal with a very noisy environment (including counter frame, reference numbers, sealing wires, etc.). Processing is performed in the following order: image input; label identification and recognition; identification of images of single counters; segmentation and normalization; measurement of the distances from the classes of archetypes $0, 1, \ldots\ldots, 9$ for each image digit; definition of the reading using also contextual logic relying on reading forecasts.

The final series of acceptance tests of this system were successfully effected in March 1980.

4.2 Application under development

4.2.1 Speech Recognition

The project aims at recognition of words and at recognition of continuous speech.

In the recognition of words, an original approach has been followed [13] to avoid the utterance inconvenience frequently present in systems working on this problem (pause between words, training of the speaker, limited vocabulary and syntax). At present an on-line system is under development, allowing the guiding of a simulated robot on a CRT display. The system should work even in a very noisy environment. The results so far obtained show that the system can recognize a fair percentage of words, uttered by either male or female untrained speakers.

The final goal of the project is the development of a system for real time continuous speech recognition of a large subset of the Italian language within general semantic constraints.

---

* **M**ulti **A**pplication **O**ptical **R**ecognition **S**ystem.

4.3 Future developments

Various applications are being studied, both in the field of pattern recognition (for example, reading of bubble chamber photograms) and in the field of process control (for example, control application in the steel industry) in which it should be possible to take advantage of EMMA's intrinsic reliability and computing power to carry out tasks with complex I/O and stringent time processing requirements.

5. CONCLUSIONS

EMMA is an industrial product that ELSAG has manufactured so far in various configurations totalling in 1980: 25 EMMA systems and 1400 AU's (processors).

EMMA was designed as a flexible tool for real time applications. The simplicity and efficiency of its structure ensure that the number of processors for a particular application is dictated only by the project specifications and not by interconnection requirements. The validity of its architecture is confirmed by the design solutions adopted, for instance, by Cm★; the two systems show a remarkable resemblance of conception even though they exploit different technologies (EMMA's project was started in 1972, Cm★'s in 1975).

EMMA has experimentally demonstrated its capabilities in the field, and may be the basis for extending the application of general purpose computing systems to those areas where so far mostly specialized hardware has been used.

6. ADDENDA

Since the time of writing two important events, regarding the EMMA multi-processor, have taken place:
  (a) SARI has been chosen by the U.S. Postal Service Administration for its program of automation. For this purpose 126 SARI systems will be installed across the U.S.A. in the next 3 years;
  (b) a MAORS system for automatic telephone counters reading will be supplied to the COMPAÑIA TELEFONICA NACIONAL DE ESPAÑA.

REFERENCES

] Bertora, F., Boccalini, C. and DiManzo, M. (1978). "Effect of memory interference in tree-structured multiprocessor systems". In *Large scale integration*, Lawson, W. W., Berndt, H. and Hermanson, G. (eds). EUROMICRO 1978, North Holland, pp. 274–278.

] Brinch Hansen, P. (1978). "Distributed processes: a concurrent programming concept". *Comm. ACM* 21, 934–941.

] Brinch Hansen, P. (1977). *The architecture of concurrent programs*. Prentice Hall, Englewood Cliffs.

] Dijkstra, E. W. (1968). "Cooperating sequential processes". In *Programming languages*, Genuys, F. (ed.). Academic Press, London and New York, pp. 43–112.

] Enslow, P. H. (1977). "Multiprocessor organization—a survey". *Comp. Surveys* 9, 103–129.

] Flynn, M. J. (1972). "Some computer organizations and their effectiveness". Trans. IEEE on Comp. C-21, 948–960.

] Fuller, S. H., Ousterhout, J. K., Raskin, L., Rubinfeld, P. I., Sindhu, P. J. and Swan, R. (1978). "Multimicroprocessors: an overview and working example". Proc. IEEE 66, 216–228.

] Hoare, C. A. R. (1974). "MONITORS: an operating system structuring concept". *Comm. ACM* 17, 549–557.

] Hoener, S. and Roehder, W. (1977). "Efficiency of a multimicroprocessor system with timeshared buses". Proc. 3rd EUROMICRO Symp., Amsterdam, pp. 35–42.

] Jensen, E. D. (1977). "The Honeywell experimental distributed processor". AICA Cong., Pisa, Italy.

] Madnick, S. E. and Donovan, J. J. (1974). *Operating Systems*. McGraw-Hill, New York.

] Peterson, J. L. (1977). "Petri nets". *Comp. Surveys* 9, 223–252.

] Rossi, C. and Stringa, L. (1977). "A linguistic approach to speech recognition via a minicomputer net". IEEE Int. Conf. on Acoustic Speech and Sig. Proc. Hartford, Connecticut.

] Stringa, L. (1977). "EMMA: Elaboratore Multi Mini Associativo". AICA Cong., Pisa, Italy.

] Stringa, L. (1978). "EMMA: An unbounded modular multi-mini processor structure". ACM Comp. Sci. Conf., Detroit.

] Stringa, L. (1974). "Hand and machine printed address reader implemented on a multi-mini based associative processor". Proc. 2nd IJCPR, Copenhagen, Denmark.

] Stringa, L. (1978). "Hand and machine printed address recognition via EMMA net". Proc. COMPCON 1978—16th Comp. Soc. Conf., S. Francisco.

] Stringa, L. and Violino, E. (1978). "Fault tolerance through a graceful degradation strategy on EMMA Multiprocessor". Rep. ELSAG ERI/029/A.

] Swan, R. J., Fuller, S. H. and Siewiorek, D. P. (1977). "Cm*—a modular multi-microprocessor". AFIPS Conf. Proc. 46, 637–644.

] Thurber, K. J. and Wald, L. D. (1975). "Associative and parallel processors". *Comp. Surveys* 7, 215–255.

] Wulf, W. A. and Bell, C. A. (1972). "C.mmp: A multi-mini processor". AFIPS Conf. Proc., 41, 767–777.

Chapter Eighteen

A Parallel Heterarchical Machine for High-level Language Processing

Adolfo Guzmán

1. INTRODUCTION AND PROJECT STATUS

This chapter presents the architecture of a parallel general purpose computer that has LISP as its main programming language. It is built of several dozens of microprocessors (Z-80's), each of them executing a part of the program.

1.1 Goals

The goals of the Project AHR (Arquitecturas Heterárquicas Reconfigurables) are:

(a) to explore new ways to perform parallel processing;
(b) to have a machine in which it will be possible to develop parallel processing languages and software;
(c) to have a tool for students to learn and practice parallel concepts in hardware and software.

1.2 Project status

Version 0 [3] of the machine has been designed and simulated. This produced Version 1 [12], which was simulated using SIMULA.

We are building Version 1 of the machine, expected to be operational [5] in the first quarter of 1981. Subsequently, a faster version will be built, possibly incorporating changes and ideas sprung from our experience with the first machine. Finally, this fast version will be used to try to attain the goals mentioned above.

About six people are involved in the project full-time. The expected uses of the machine also include picture processing, finite element methods, engineering calculations, and distributed processing.

1.3 Main features

The AHR machine has the following characteristics:
(a) general purpose;
(b) parallel processor;
(c) heterarchical. It means that there is no hierarchy among the processors; there is no "master" processor, or controller. All the processors are at the same level;
(d) asynchronous operation;
(e) it has LISP as its main programming language;
(f) processors do not communicate directly amongst themselves. They only "leave work" for somebody else to do it;
(g) no input/output. This is handled by a minicomputer to which the AHR machine is attached;
(h) no operating system (software). Most of the LISP operations, as well as the garbage collector, are written in Z-80 machine language;
(i) the AHR machine works as a slave of a general purpose computer (a mini);
(j) gradually expandable. More microprocessors can be added as additional computing power is needed [9].

1.4 Funtional notation

The AHR machine obtains its parallelism by parallel evaluation of the arguments of functions. For instance, in f(a,b, g(u,g(x,b))), first x and b are evaluated; then g of them, in parallel with u; then g of the result, in parallel with a and b. That is, evaluation occurs from bottom up, or from the inside to the outside of the expression. This is in accordance with the rule for evaluation of a function: "to evaluate a function, the arguments have to be already evaluated".

Recursion is handled [3] by substituting the function name ("FACTORIAL") by its function definition (LAMBDA (N) (IF (EQ N O) 1 . .)) when evaluating it.

The machine works with pure LISP, without SETQ's, GOTO's, Label's, RPLACA.

2. THE PARTS OF THE AHR MACHINE

In this section the constituents of the machine are described; section 3 explains how the machine works. Refer to Fig. 2—"The AHR machine".

2.1 Passive memory

This memory holds lists and atoms; it holds partial results and parts of programs that are not being executed at the moment.

Originally, the programs to be executed reside here, and they are copied to the grill (see below) for their execution. As new data structures are built as partial results of the evaluation, they come to the passive memory to reside.

2.2 The grill

This memory holds the programs that are being executed. A program, once in the grill, is being transformed into *results*, as the result of its evaluation.

Programs reside in the grill in the form of *nodes* (see Fig. 1). Each node is pointed at by its sons (its arguments), and its *nane* field contains the number of non-evaluated arguments. Nodes with nane = 0 are ready for evaluation.

2.3 The LISP processors

These active units are microprocessors (about several dozens of Z-80's) that obtain from the grill nodes ready for evaluation, and, after evaluation, return *results* (s-expressions) to the grill. Each LISP processor knows how to execute every LISP primitive. Each of them works asynchronously, without communicating with other processors.

The processors obtain new work to be done from the distributor, through the *high speed bus*. This work comes as a node ready to be evaluated.

Only nodes with nane = 0 come up to the LISP processors for evaluation. So, for instance (CAR '(A B C)') will evaluate to A. The node (CAR '(A B C)') has become the result A. The LISP processors has to do, after evaluation, the following things:

(1) Insert the new result A in the cell (in the grill) pointed to by the node (CAR '(A B C)'). That is, insert such result in a slot of the father of the evaluated node (see such slots in Fig. 1).

(2) Release the grill space occupied by node (CAR '(A B C)').

(3) Subtract 1 from the nane of the father.

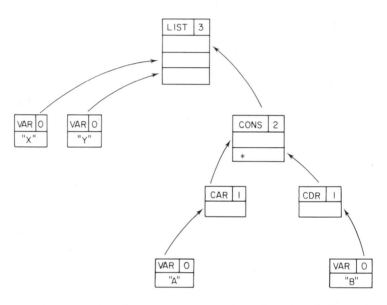

Fig. 1. Nodes in the grill. (above) the LISP expression to be evaluated. (below) how it is structured into nodes, each node being a function or a variable. Each node shows a number: its nane, or number of non-evaluated arguments. When a node has a nane of zero, it means that node is ready for evaluation.

Empty words are slots where the results of evaluation will be inserted. For instance, the results of (CDR B) will be inserted in the slot marked "★".

(4) If the new nane (of the father) is zero, inscribe the father in the FIFO: the father is now ready for evaluation.

These steps are done by the processor simply by signalling to the distributor that it has finished, and that its results should be handled in mode "normal end" (burocracia de salida, in Spanish [12]); the distributor itself performs the requested steps.

Notice that in this form nobody has to search the grill looking for nodes with nane = 0, because as soon as they appear, they are inserted into the tail of the FIFO.

The LISP processors have access to the passive memory (where lists and atoms reside), and to the variable memory, where we have the vlaues of variables. A LISP processor is either *busy* (evaluating a node) or it is *ready* to accept more work (another node).

2.3.1 The high speed bus

Connecting each LISP processor with the distributor is a high speed bus that goes into the private memory of each processor. The new node that the distributor throws is inserted (through the high speed bus) into the memory of the selected processor. Then, the processor is signalled to proceed.

2.3.2 The slow speed bus

This bus runs from the I/O processor (the mini to which the AHR machine is connected) to each box. It is not shown in the diagrams, nor it is explained furthermore in this article [see 5]. Through this bus each processor is loaded with programs, prior to starting the machine. Also, in the debugging stage, the slow bus is used to pass statistical information to the I/O processor.

2.4 Variable memory

This memory contains pairs of (variable, values), and it is organized as a tree, or a collection of a-lists, where each pair of (variable, values) points to older pairs. It is accessed by the LISP processors, and it is augmented (a branch of the tree grows) after each LAMBDA binding.

2.5 The distributor

This piece of hardware communicates the grill with the LISP processors. The distributor keeps in the *FIFO* (a memory) and array of nodes ready to be evaluated; these nodes are thrown, one in each cycle of the distributor, to the LISP processors that are ready to accept new work. An *arbiter* decides which LISP processor obtains the node; an exchange is done (through the high speed bus) between that LISP processor and the distributor, the processor accepting the node and releasing the result of the previous evaluation. The distributor stores such results in the grill, in the address indicated within the result. Generally, this result is stored in a slot of the node which is father of the node just evaluated. An overall view of the machine is shown in Fig. 2.

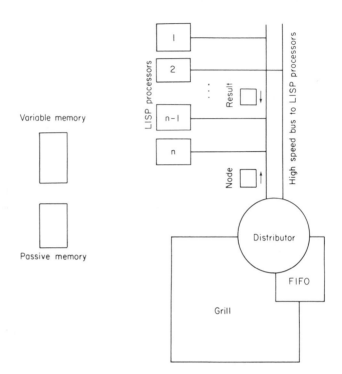

Fig. 2. The AHR machine. LISP processor 2 is ready to accept more work. The distributor fetches a node (to be evaluated) from the FIFO and sends it to processor 2, while accepting the results of the previous evaluation performed by such a processor. That result is stored in the grill, in a place indicated in the destination address of the result.

Such exchange of new work-previous result is performed at each cycle of the distributor.

Version 2 of the AHR machine will gain speed over version 1, mainly by building a fast distributor.

The LISP processors also have access (connections not shown) to the variable and passive memories.

2.5.1 The FIFO

The FIFO is a first in-first out memory that holds pointers to nodes (in the grill) ready to be evaluated. The distributor fetches such nodes through the head of the FIFO, while new nodes to be evaluated are inserted through its tail [5].

2.5.2. The arbiter

If several LISP processors become ready to accept more work, the arbiter (a hardware) selects one of them, which will receive the node thrown by the distributor. If every processor is busy, the cycle of the distributor is wasted, since no processor accepts the node that the distributor is offering.

2.6 The I/O processor

It has been said that the AHR machine can be seen as a peripheral of a general purpose minicomputer. But this mini can also be considered as a peripheral of the AHR machine; we thus talk of such a mini as the I/O processor.

Input/ouput will be described in the next section.

3. HOW THE MACHINE WORKS

3.1 Input

The user sits at a terminal of the mini (I/O processor) which is master of the AHR machine. He uses a common editor, discs and the normal operating system of the mini. When he is ready to run a program, he loads it from disc into a part of the address space of the mini which is really the passive memory of the AHR machine (see Fig. 3). In this way, the program is loaded (already as list cells) in the passive memory. A signal from the I/O processor to the AHR machine signifies that execution should begin. Together with this signal an address is passed, indicating where in passive memory the program to be evaluated resides.

3.2 Starting

It is assumed that each LISP processor already has its programs loaded in its private memory.

When the AHR machine receives the "start" signal, the distributor throws a node (called the RUN node) to the LISP processor. This node points to the program which will start.

The program (in passive memory) is copied by more and more LISP processors (the more leaves or branches a program has, the more processors help to copy it) into the grill. Nodes with nane = 0 are inserted by the LISP

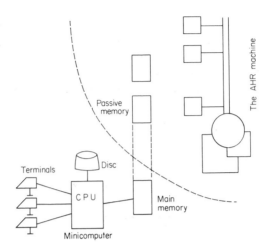

Fig. 3. The AHR machine as a slave. The AHR computer is shown as another peripheral of a general purpose minicomputer. The address space of the mini comprises the passive memory of AHR, through a movable window of 4 K addresses.

processors into the FIFO, so that some other LISP processor with execute them. Finally, the program has been copied into the grill. Notice that at the same time of copying, some nodes with nane = 0 could have been evaluated by some other LISP processors.

3.3 Evaluation

When a LISP processor is idle, it signals to the distributor, meaning that it is ready to accept more work.

The distributor chooses (with the help of an arbiter) one of several idle processors, and through the high speed bus it injects a new node (taken from the grill through the head of the FIFO) into its private memory. It then signals such a processor to start.

The LISP processor "discovers" the node in its own memory, with all the arguments already evaluated. The LISP processor proceeds to perform the evaluation that the node demands. Suppose it is LIST, and its arguments are (A B), M and N. It then has to address the passive memory in the mode "give a new cell". Such a cell is given by a cell dispatcher (hardware attached to passive memory). Three new cells have to be requested. Then the LISP processor forms the result: ((A B) M N). For this, it has to store pointers to (A B), to M and to N, into passive memory, in the new cells already obtained. Then, it stores the result (which is a pointer to passive memory) into a special place ("results place") of its private memory. It has finished. It signals to the distributor that it is ready to accept more work. The distributor will insert new work (another node with nane = 0) into the private memory of the processor, but it will also collect (through the high speed bus; see Fig. 2) from the "results place" in private memory, the result ((A B) M N). The distributor will store this result into a slot in a node in the grill. The address in the grill of this slot was known to the (LIST (A B) M N) node, because each node points to its father. Thus, the distributor has no problem in finding where to store the result: such address is found also in the "results place", together with the result ((A B) M N).

The distributor has to do one more thing: it has to substract one from the nane of the father (which has just received the result ((A B) M N)). And if such nane becomes zero, then a pointer to the father is insterted by the distributor into the FIFO through its tail. Finally, the distributor has to free the cell of the node (LIST (A B) M N), so that this grill space could be re-used [10].

The distributor is very fast compared with the speed of the LISP processor. This will be even more true if we code "complicated" LISP functions (such as MEMBER OR FACTORIAL) in Z-80 machine language, instead of "simple" LISP functions, such as CDR.

Due to such differences in speed, the distributor can keep many LISP processors working; if the distributor is 100 times faster than the (average) LISP function, it could keep 100 LISP processors functioning. It pays to make a fast distributor.

3.4 Output

Finally, the whole program has been converted into a single result (let us say, a list) deposited in passive memory. The AHR machine now signals the mini (or I/O processor), giving it also the address in passive memory where the result lays. The mini now accesses the passive memory as if it were part of its own memory (remember, their address spaces overlap), and proceeds to the (serial) printing process. Execution has now finished.

4. HARDWARE CONSIDERATIONS

4.1 LISP processors

The first version of the machine will have 5 LISP processors, and the "mini" or I/O processor is another Z-80. Each LISP processor will have 4 K bytes of private memory, where a pure-LISP interpreter will reside [8].

The maximum number of LISP processors is 64. It could be increased further, but a new arbiter needs to be designed in that case.

4.1.1 The high speed bus

The distributor inserts a node (7 words of 32 bits) into the private address space of the selected LISP processor, through the high speed bus. It does this in 0.5 μs. It runs from the distributor to all LISP processors. It carries nodes and results.

4.1.2 The low speed bus

In a 16 bits low speed bus, 8 of the bits indicate which LISP processor is addressed, the other 8 bits carry data. It runs from the I/O processor to the LISP processors.

An additional use of the low speed bus is to broadcast to the LISP processors the number of a program that needs to be stopped or aborted.

4.2 Passive memory

It consists of up to 1024 K words of 22 bits; it contains the input ports, list space, output ports and atom space. Version 1 will have only 64 K words. Access time is 150 ns. It has a parity bit.

4.3 The grill

It consists of up to 512 K words of 32 bits. It is divided logically in nodes, each with 7 words. Version 1 will have 8 K words. Access time is 55 ns. The grill contains the nodes that are about to be evaluated.

4.4 Variable memory

It consists of up to 512 K words of 32 bits. This memory contains names of variables and their values at a given time. The variable memory also contains real numbers, in its lower half. In its upper half it has "environments", which are lists of cells of 5 words each. Version 1 will have 16 K words. Access time is 150 ns.

4.5 The distributor

The distributor passes nodes from the grill to the LISP processors, and stores in the grill the results coming from the LISP processors. There are two versions of the distributor.

4.5.1 First version of the distributor

This first version [10] is implemented through a Z-80, using a program that performs all the functions of the distributor. It runs slowly, in the sense that it distributes nodes at low speed. It is further described in section 5.

4.5.2 Second version: fast distributor

This version is not yet built; it will become part of version 1 of the machine. It will be built either from bit-slice microprocessors, or from PAL's.

4.5.3 The FIFO

Of a maximum of 512 K words of 19 bits, the FIFO contains pointers to the nodes in the grill. Version 1 will be of 4 K words. Its access time is 55 ns.

4.5.4 The arbiter

There are really three arbiters—for passive memory, variable memory and for the grill. Each arbiter takes 400 ns to respond, and it may handle up to 64 processors. Each processor has a fixed priority, varying from 1 to 64. Each processor has a different (unique) priority.

4.6 The I/O processor

This is actually built around a Z-80 that works as a general purpose computer. Its main functions are to:
  (a) talk to the users; to read their input and to print their results;
  (b) store user files in its disc;
  (c) initialize the AHR machine;
  (d) load into passive memory, through the window, the programs loaded from disc;
  (e) begin garbage collection;
  (f) end garbage collection;
  (g) Actually, the garbage collector runs in the I/O processor.

5. SOFTWARE CONSIDERATIONS

5.1 The LISP interpreter

A LISP interpreter runs in each LISP processor. It interprets pure LISP (only evaluations; no SETQs, RPLACDs or other operators). The garbage collection is not done by the LISP processors at this moment.

For the first version, the LISP interpreter will do argument checking of the LISP functions. This will remain as an option in the second version of the AHR machine.

5.2 The garbage collector

For the first version of the machine, this will be a "normal" serial garbage collector, running in the I/O processor. While it is working, the LISP processors remain idle. For the second version, it will be a parallel incremental garbage collector, running in the LISP processors.

Garbage collection is done for passive memory (list cells) and for the real numbers region of variable memory (where it compactifies memory). In the "environments" zone of variable memory and in the grill (nodes), there is no need to re-collect garbage, because used space, as soon as it is abandoned in these two places, is inserted (by hardware) into a list of free environment cells (for variable memory) or into a list of free nodes (for the grill).

5.3 The distributor (first version)

This is a piece of software [10] running in a Z-80, that emulates all the functions that the "real" (hardware) distributor performs. It is slow in this sense, but it is flexible and helps in the debugging of the AHR machine; it may be run "step-by-step" to see the flow of information. It also keeps statistics of use of hardware and software.

5.4 Editing

Editing of LISP programs is done outside the AHR machine, using the operating system and editor of the I/O processor. After editing, the program is filed on disc. From here, a loader (running in the I/O processor) converts it into list cells and brings the program to passive memory (see Fig. 3).

6. RELATED WORK AND MACHINES

6.1 Greenblatt's LISP machine

This is a single processor machine [14] built for high speed LISP comput-ations. It does not pretend to be an experiment in parallel hardware; it gains its speed and power from careful design of the software and machine architecture, as well as from the experience of the builders with the LISP language.

6.2 Parallel LISP machine

This machine [7] is a loosely coupled multiprocessor for applicative languages such as LISP. It is the machine most closely resembling ours, in its application.

6.3 Data flow machines

These machines [13] resemble the AHR architecture in that data is directed through "boxes" that process them. The flow of executions is controlled, like in our design, by what previous results are ready (available). The cited article describes a machine that uses different colours of tokens to mark "this result", "previous result", and so on.

6.4 ZMOB

A collection of Z-80's around a conveyor belt, this machine [11] may be applied to image processing and numerical calculations. Each microprocessor has its own private memory. They do not have direct access to a common memory (as AHR does), but behind one of the micros, a huge central memory or mass memory may reside.

6.5 PM4

This is a machine [2] suitable for image processing. It is a dynamically reconfigurable multimicroprocessor-based machine. It can be partitioned into several groups of processors which may be assigned to execute multiple independent SIMD processes and MIMD processes.

6.6 The language "L" for image processing

"L" is a language suitable for processing of images. It is mentioned here because it may be implemented in a parallel machine [4], such as the AHR computer. The language is described elsewhere [1] in this book. It was designed mainly as a result of our experience in picture processing of multi-spectral images [6]. "L" has not been implemented.

7. CONCLUSIONS

The architecture of the AHR computer shows that it is possible to build a multiprocessor of the MIMD type, where each processor does not explicitly communicate with other processors. In the AHR design, a processor does not know how many other processors are there, or what they are doing. It is not possible to address a processor: "here I have a message for processor number 4."

The construction of new software has been kept low by connecting the machine to a general purpose computer, thus being able to use already available operating systems for time sharing, text editors and loaders.

Once the machine is built, experimentation will begin in the design of parallel languages and ways to express "powerful" commands in heterarchical fashion. Also, if the amount of access to memories for each processor is low, it may be possible to place each micro in a remote place, thus achieving some class of distributed computing. That is, a micro can process local work (through Basic, for instance) as well as remote (LISP) work.

Finally, the AHR machine shows how it is possible to design a heterarchical system, where none of the processors tells the others what to do, in what order to do it, or what resources are available to whom.

ACKNOWLEDGEMENTS

The AHR machine is being built by the members of the AHR Project, to whom I express my appreciation for their time, effort and enthusiasm.

Work herein described has been partially supported by Grant 1632 from CONACYT, the National Council for Science and Technology (Mexico).

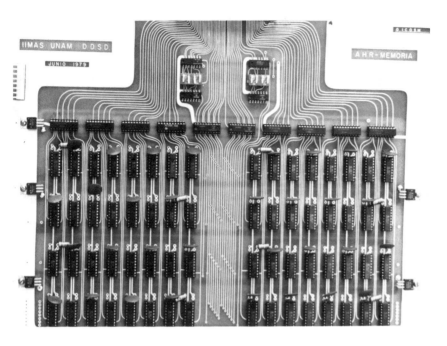

Fig. 4. The FIFO. This figure shows the FIFO memory that stores pointers to the grill, containing nodes ready to be evaluated.

REFERENCES

Barrera, R., Guzmán, A., Ginich, A. and Radhakrishan, T. (1981). "Design of a high level language for image processing". In *Languages and architectures for image processing*. M. J. B. Duff and S. Levialdi (eds). Academic Press, London and New York.

Briggs, F. A., Fu, K. S., Hwang, K. and Patel, J. H. (1979). "PM4: A reconfigurable multiprocessor system for pattern recognition and image processing". Tech. Rep. TR-EE-79-11, School of Electr. Eng., Purdue University.

Guzmán, A. and Segovia, R. (1976). "A parallel reconfigurable LISP machine". Proc. Int. Conf. on Inf. Sci. and Syst., University of Patras, Greece, pp. 207–211.

Guzmán, A. (1979). "Heterarchical architectures for parallel processing of digital images". Tech. Rep. AHR-79-3, IIMAS, Nat. Univ. of Mexico.

Guzmán, A., Lyons, L. *et al.* (1980). "The AHR computer: construction of a multi-processor with LISP as its main language" (in Spanish). Tech. Rep. AHR-80-10, IIMAS, Nat. Univ. of Mexico.

Guzmán, A., Seco, R. and Sanchez, V. (1976). "Computer analysis of LANDSAT images for crop identification in Mexico". Proc. Int. Conf. on Inf. Sci. and Syst., University of Patras, Greece, pp. 361–366.

Adolfo Guzmán

[7] Keller, R. M., Lindstrom. G. and Patil, S. (1979). "A loosely-coupled applicative multi-processing system". AFIPS Conf. Proc. **48,** 613–622.

[8] Norkin, K. and Gomez, D. (1979). "A new description for data transformations in the AHR computer". Tech. Rep. AHR-79-4, IIMAS, Nat. Univ. of Mexico.

[9] Norkin, K. and Rosenblueth, D. (1979). "Towards optimization in AHR". Tech. Rep. AHR-79-5, IIMAS, Nat. Univ. of Mexico.

[10] Penarrieta, L. (1980). "Error detection in the AHR computer" (in Spanish). Tech. Rep. AHR-80-9, IIMAS, Nat. Univ. of Mexico.

[11] Rieger, C., Bane, J. and Trigg, R. (1980). "ZMOB: a highly parallel multiprocessor". Tech. Rep. TR-911, Dept. of Comp. Sci., University of Maryland.

[12] Rosenbleuth, D. and Velarde, C. (1979). "The AHR machine for parallel processing: 1st stage" (in Spanish). Tech. Rep. AHR-79-2, IIMAS, Nat. Univ. of Mexico.

[13] Watson, I. and Gurd, J. (1979). "A prototype dataflow computer with token labeling". AFIPS Conf. Proc. **48,** 623–628.

[14] Weinreb, D. and Moon, D. (1979). *LISP machine manual.* M.I.T.A.I. Lab., Cambridge, Massachusetts.

Chapter Nineteen

FLIP: A Multiprocessor System for Image Processing

P. Gemmar, H. Ischen and K. Luetjen

1. INTRODUCTION

The research and development of computerized image processing has brought digital image processing a more and more growing field of applications. But there are increasin_ difficulties for practical applications and even for experimental simulations, since the complex algorithms are time consuming and image data to be processed have increased both in terms of image size and number of images.

In many cases the architecture of image processing systems is influenced by the structural requirements of image processing tasks. The evaluation of those requirements shows 'iat different processing structures are needed for the various steps of image processing. With FLIP a flexible multiprocessor system has been developed which combines simultaneous processing and structural flexibility. By this means the FLIP system provides a suitable processing power and computing efficiency. FLIP's typical application is to perform homogeneous operations which can mainly be found in the fields of picture preprocessing and feature extraction, e.g. image filtering (linear, non-linear), contrast sharpening, contour enhancement, local image convolution, and local image correlations, etc.

2. CONSIDERATIONS ON EFFICIENT PROCESSING TECHNIQUES

The parallelism of an operation is usually not exploited in practice. The programmer typically handles a parallel problem by a topological transformation stretching the job into a suitable form for a one-at-a-time hardware. The processing of a job requires occupation of equipment space over a certain length of time and can be represented as an enclosed area in a space-time diagram [3]. If parallel hardware is available, the execution of a job involves sweeping the job area by a parallel mechanism to produce complete coverage. Both parallel and pipelined hardware are best in handling parallel jobs. Computing efficiency of parallel hardware can be defined as:

$$E = \frac{\text{total space time of job}}{\text{total area swept by the multiprocessor}} \qquad (1)$$

To illustrate the design aspects and structural requirements leading to a new processing structure the following example of a simple algorithm called "stroke difference" will be given. This algorithm belongs to the above mentioned class of parallel image processing operations and can be used to generate the first derivate of an image F. With the submatrix notation of Fig. 1 the stroke difference is given by (2).

To evaluate the structure of a given mathematical expression it is useful to draw a corresponding computing graph. For this purpose the algorithm is decomposed into its elementary operations. The computing graph in Fig. 2 displays the elementary operations by its nodes and the processing flow by connecting arrows.

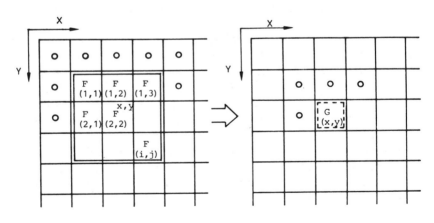

INPUT IMAGE F   OUTPUT IMAGE G

Fig. 1. Submatrix notation for a 3 × 3 window

$$Gx(x,y) = |(F(1,1)+F(2,1)+F(3,1) - (F(1,3)+F(2,3)+F(3,3))|/3$$
$$Gx(x,y) = |(F(1,1)+F(1,2)+F(1,3) - (F(3,1)+F(3,2)+F(3,3))|/3$$
$$G(x,y) = (Gx(x,y) + Gy(x,y))/2 \qquad (2)$$

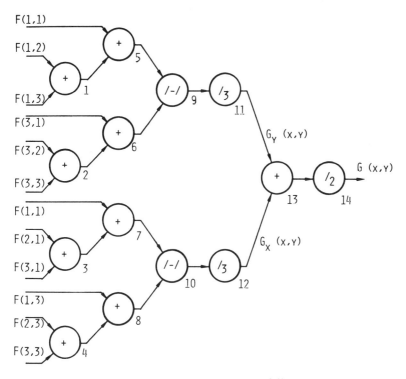

Fig. 2. Computing graph for stroke difference.

This graph shows two specific principles to speed up computation:
  (i) Parallel Processing: this type of processing is characterized by the concurrent computation of unrelated operations (e.g. operations 1–4 in the first stage, and 5–8 in the second stage, etc.).
 (ii) Pipelining: the pipelining of operations is characterized by the sequential flow of data through the processors and the parallel working of all stages (e.g. operations 1, 5, 9, 11, 13 and 14, etc.).
The speed of data flow in this structure is dictated by the slowest operation (or process) within a pipeline. The processing structure depicted in Fig. 2 is called a cascade and its operation can briefly be described as simultaneous processing.

Cascading of operations becomes more difficult if jobs are getting more complex and processors must execute a number of operations (program) instead of remaining simple operators (e.g. adders, etc.). To obtain good performance it is necessary to order the operations in a proper way and to distribute equal loads for each processor within a cascade. Nevertheless, even if job parallelism can fully be exploited by the hardware the synchronization and control of processes is difficult to maintain. Consequently, it is not meaningful to control the system from one central source.

The processing cascade as depicted in Fig. 2 looks somewhat like a funnel-shaped processing scheme. For fast and efficient computation a suitable flow of the data is required. Figure 3 shows the flow of the data for this type of processing.

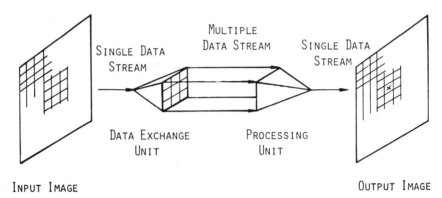

Fig. 3. Data flow with window operations.

The consideration of the above mentioned structural requirements to an image processing system directly led to the design of FLIP. As a result FLIP comprises the following hardware features:

(i) Multiprocessor system: high computation power is provided by simultaneous operation of several processing elements (16 individual processors).

(ii) Structural programmability: it is possible to arrange the processors in nearly any wanted processing structure by software means.

(iii) Parallel data input stream: a programmable data source (PEP) provides a high capacity parallel data input stream.

(iv) Effective and simple synchronization: the synchronization of individual processing elements is supported by hardware (asynchronous data-flow controlled mode of operation).

(v) Programmable operations: complex and different operations can be performed in each processor, as each processor has its own program memory. Both the processing properties (operation code) and the configuration of processors (operand addresses) are established by the programs of the individual processors.

3. FLIP SYSTEM CONFIGURATION

The FLIP system operates as a peripheral device in conjunction with a host computer (Fig. 4). FLIP consists of a central processing unit FIP (Flexible Individual Processors) and a data exchange processor PEP (Peripheral data Exchange Processor). The FIP is built with 16 IPs (Individual Processor). The PEP, which is mainly a fast buffer device, comprises three internal processors to provide the high data rate required by FIP. Additionally, FLIP is directly connected to a MOS image memory (768 Kbyte) via a high speed data path. Beside several mass storage devices (disc, tape) and image input/ouput devices the host provides two additional links to another minicomputer (PDP 11/70 with further image input/output devices) and VAX 11/780 (4 Mbyte main storage) general purpose computer, respectively.

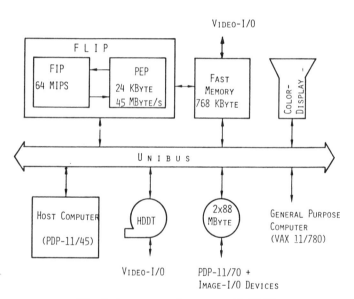

Fig. 4. Image processing system with FLIP.

3.1 Processing unit FIP

An important feature of the realized FIP system is its flexible and pro-grammable structure. This gives the possibility to arrange the ensemble of 16 individual processors (IP1 to IP16) according to the topology of the processing task. To limit the implementation effort and to improve the computing speed the system is designed to work without any system clock. A self-synchronizing scheme achieved by the data flow is used for sequence control. For this reason a bus system is built, which physically connects each IP to all the others (Fig. 5). In order to ease programming of FLIP and to increase system speed each processor is provided with two independent input ports and one output port. By this way, each individual processor may have two data inputs and one data output on different buses at the same time (three address machine).

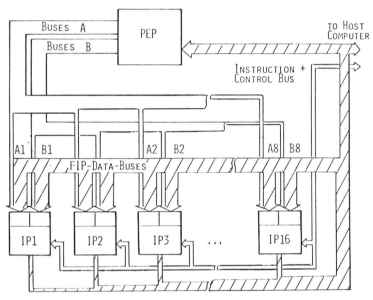

Fig. 5. The FIP bus system.

All individual processors of FIP are identical. To increase the processing speed, data and instructions are internally separated both in storage and signal path. Storage room provided is 50 bytes (8 bit) for data and 256 words (32 bit) for programs. Figure 6 shows the functional block diagram of an IP.

The instruction repertoire of each IP includes the basic arithmetical and logical instructions performed on 8-bit data (e.g. add, substract, negate, or bit clear, etc.), the 8-bit by 8-bit multiplication, shift instructions, branches and subroutine jumps, and instructions to handle multiple precision data. The execution time of each instruction is dependent on the complexity of the applied execution and addressing mode. The measured mean execution time is about 250 ns. The IP's are built with LP-Schottky-TTL SSI and MSI circuits.

Using Flynn's [5] classification approach for computer organizations FLIP can be considered a MIMD (multiple instruction multiple data)—multiprocessor system. Since each processor (IP) has its own program memory it is possible to let it perform complete programs. This may be advantageous in such cases, where algorithms to be computed are less complex or cannot be decomposed in operations or subtasks of equal load for each IP involved. Then a more efficient usage of FLIP often will be achieved by running the system as a "pseudo"-SIMD (single instruction multiple data)—machine with all IP's performing the same program. Due to FLIP's structural flexibility this kind of parallel processing is fully supported by hardware and does not mean any expense for the user.

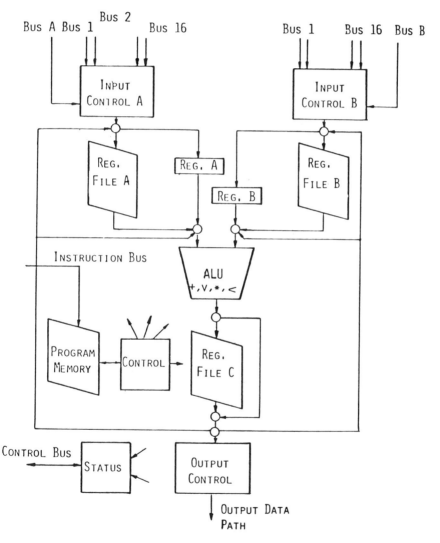

Fig. 6. *Structure of an Individual Processor (IP).*

3.2 Data exchange unit PEP

The data flow between FLIP and the host computer is accomplished by three exchange processors (EP), a data memory, and associated interface, respectively, all comprised within PEP (Fig. 7). PEP is designed to provide input data for the FIP. A multiple data stream is formed from the single input data stream coming from the host system. Working on a 5*5 submatrix, for example, data rates to FIP are required 25 times the rate of the sequential input data stream. To achieve a suitable data rate PEP and FIP are connected together by 16 independent buses A and B providing a maximum data rate of 45 Mbyte/s.

The access to the data stored in the data memory is free programmable and almost unlimited. Additional addressing hardware is available for the homogeneous type of image processing, so that the user is not concerned with the special problems for this type of data input and output. FLIP acts like a

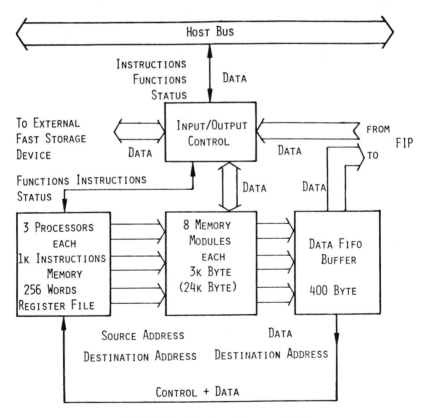

Fig. 7. Data Exchange Processor PEP.

peripheral device controlled and supervised by the host computer. Therefore, within the host a program must provide at least two independent data streams:

(i) One input data stream usually originating from a mass storage device (e.g. disc).

(ii) One output data stream going to a mass storage device or a display device.

The input data stream to FLIP is handled by the Input/Output Controller and transferred to the PEP data memory. The multiple data stream to the FIP is controlled by the three PEP processors. The data are stored in the data buffers and transferred to the FIP processors. FIP provides three output buses to return results. The results are transferred to the host, the data memory, or the MOS-image memory, respectively, under the contol of the Input/Output Controller.

4. FLIP—PRINCIPLE OF OPERATION

To establish a desired processing structure on FLIP, the individual processors of FIP as well as the PEP processors (EP1–EP3) must be programmed. As soon as all programs are loaded the processing structure is latent in the system and processing can be started. This is usually done by delivering the image data to the FLIP. This can be achieved by DMA (Direct Memory Access) data transfer initiated by the host computer. This data delivery scheme provides the host computer full control over the entire processing cycle.

Nearly any desired processing structure of FIP processors can be established from the pipeline over the cascade up to other parallel structures. Two simple examples are illustrated in Fig. 8.

The only means to logically connect two FIP-processors is by the address parts (operand address = bus address) within the instructions of the processor requesting the data. Data paths are switched while performing instructions with (external) bus operands. Each of these input operands requires an operation with output to the bus at the addressed IP. Data transmissions between two IPs are controlled by a hand-shake mechanism. In the case of an input operation without valid data at one or both input ports, respectively, the processor will remain in a waiting state.

Once an instruction containing the addressing mode "data to output-bus" is executed data output will be initiated and program execution be resumed. If the last output data was not taken by another processor the execution of a second instruction with output to bus has to wait until the first data transfer is accomplished. If the previous data transfer is already finished the execution proceeds without interruption. In this way, the FLIP has an asynchronous sequence control achieved by mutual synchronization of the individual processors with their joint data flow.

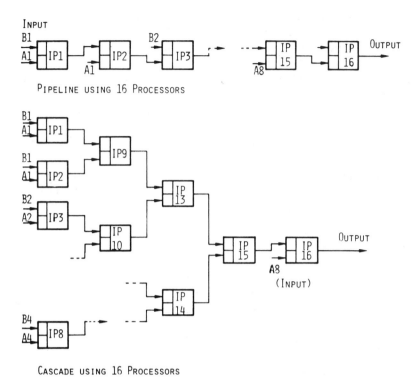

Fig. 8. Processing configurations with FIP.

4.1 Supporting software

The FLIP is programmed in its machine languages called FAL (FIP Assembly Language) and PAL (PEP Assembly Language), respectively. The object modules are linked by FLINK (FLIP LINKage editor) which arranges the relative bus addresses within the objects to establish the desired logical structure on FLIP. Program execution is initiated and supervised by the FLEX (FLIP EXecutive) utility running on the host computer. FLEX controls all steps: the transfer of image data from a mass storage to FLIP, the FLIP execution, and finally the storing of the results produced by FLIP. Additional features of the control program FELX are the capability to debug FLIP programs and to locate program errors by the use of trace routines, and to evaluate program efficiency by monitoring the activities of FLIP (FIP and PEP) by time measurements.

5. FLIP IMAGE PROCESSING APPLICATIONS

Considering the special capabilities of FLIP described in the preceding sections, one can easily realize that FLIP will find advantageous applications in a wide range of picture processing. A good deal of frequently used tasks like simple array operations for image filtering (linear or non-linear), contour sharpening, and image enchancement, etc. imply parallel processing structures to be directly implemented with FLIP. Obviously, suitable FLIP-processing structures do exist for other time consuming image processing functions, e.g. for local image correlation, image convolution or some grey level statistics, etc. In Table 1 FLIP performance data are given for a selection of image processing operations already implemented on FLIP.

Table 1
FLIP execution times for 512*512 pixels

Window size	Element diff. (s)	Stroke diff. (s)	Two-dim. conv. (s)	Median filt. (s)	Laplace filt. (s)
3*3	0.1	0.2	0.25	1	0.2
5*5	0.1	0.25	0.6	7	0.2
7*7	0.1	0.3	1.1	25	0.2

6. CONCLUSIONS

A powerful and suitable image processing system has been designed and realized. The developed system possesses a flexible multiprocessor system as central processing unit and a high speed data exchange processor for fast and convenient data input/output. The FLIP is able to perform up to 64 MIPS and can process image line lengths of up to 10 Kbyte (depending on the submatrix size) and with the number of lines unlimited.

Studies have mainly been made with homogeneous operations in the field of image preprocessing. Other applications like local image correlation or operations for feature extraction have been examined. It was found, that it takes FLIP about one second to perform most of the considered image processing tasks on images with 512*512 pixels. This means that the time required for processing is often less than the time that is needed to put in the commands for the processing system and to take the image data to and from a storage device, for example from a disc.

REFERENCES

[1] Vorgrimler, K. (1976). "Zur leistungssteigerugn von mehrprozessorsystemen fuer die verarbeitung digitaler bildinformation". Dissertation, University of Karlsruke, Federal Republic of Germany.

[2] Holdermann, F. and Kazmierczak, H. "Processing of grey-scale pictures". *Comp. Graph. and Image Proc.* **1**, 66–80.

[3] Chen, T. C. (1971). "Parallelism, pipelining and computer efficiency". *Comp. Design* pp. 69–74.

[4] Vorgrimler, K. and Gemmar, P. (1977). "Structural programming of a multiprocessor system". Proc. Int. Conf. on Parallel Computers—Parallel Mathematics, Int. Assoc. for Math. and Comp. in Simulation, Munich, Federal Republic of Germany, pp. 191–195.

[5] Flynn, M. J. (1972). "Some computer organizations and their effectiveness". Trans. IEEE on Comp. **C-21**, 948–960.

Chapter Twenty

PASM: A Reconfigurable
Multimicrocomputer System
for Image Processing

H. J. Siegel

There are several types of parallel processing systems. An SIMD (single instruction stream–multiple data stream) machine typically consists of a set of N processors, N memories, an interconnection network, and a control unit (e.g. Illiac IV). The control unit broadcasts instructions to the processors and all active ("turned on") processors execute the same instruction at the same time. Each processor executes instructions using data taken from a memory with which only it is associated. The interconnection network allows interprocessor communication. An MIMD (multiple instruction stream–multiple data stream) machine typically consists of N processors and N memories, where each processor can follow an independent instruction stream (e.g. C.mmp). As with SIMD architectures, there is a multiple data stream and an interconnection network. A *partitionable* SIMD/MIMD system is a parallel processing system which can be structured as one or more independent SIMD and/or MIMD machines. In this chapter, the basic organization of PASM, a partitionable **SIMD/MIMD** system being designed at Purdue University for image processing and pattern recognition, is briefly overviewed.

SIMD machines can be used for "local" processing of segments of images in parallel. For example, the image can be segmented, and each processor assigned a segment. Then, following the same set of instructions, such tasks as line thinning, threshold dependent operations, and gap filling can be done in

This chapter is a summary of [16]. The research was supported by the Air Force Office of Scientific Research, Air Force Systems Command, USAF, under Grant No. AFOSR-78-3581. The United States Government is authorized to reproduce and distribute reprints for Governmental purposes not withstanding any copyright notation hereon.

parallel for all segments of the image simultaneously. Also in SIMD mode, matrix arithmetic used for such tasks as statistical pattern recognition can be done efficiently. MIMD machines can be used to perform different "global" pattern recognition tasks in parallel, using multiple copies of the image or one or more shared copies. For example, in cases where the goal is to locate two or more distinct objects in an image, each object can be assigned a processor or set of processors to search for it. An SIMD/MIMD application might involve using the same set of microprocessors for preprocessing an image in SIMD mode and then doing a pattern recognition task in MIMD mode.

PASM is a special purpose, dynamically reconfigurable, large-scale multi-microprocessor system. Due to the low cost of microprocessors, computer system designers have been considering various multimicrocomputer architectures, such as [1–3, 6]. PASM was the first multimicroprocessor system in the literature to combine the following features: (i) it can be partitioned to operate as many independent SIMD and/or MIMD machines of varying sizes; and (ii) a variety of problems in image processing and pattern recognition will be used to guide the design choices.

Figure 1 is a block diagram of the basic components of PASM. The *System Control Unit* (SCU) is a conventional machine, such as a PDP-11, and is responsible for the overall co-ordination of the activities of the other components of PASM. By carefully choosing which tasks should be assigned to the SCU and which should be assigned to other system components (such as the Memory Management System), the SCU can work effectively and not become a bottleneck.

The *Parallel Computation Unit* (PCU) contains $N = 2^n$ processors, N memory modules, and an interconnection network. The PCU *processors* are microprogrammable microprocessors that perform the actual SIMD and MIMD computations. The PCU *memory modules* are used by the PCU processors for data storage in SIMD mode and both data and instruction storage in MIMD

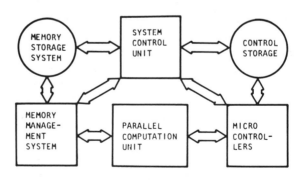

Fig. 1. Block diagram overview of PASM.

mode. A memory module is connected to each processor to form a processor–memory pair called a *processing element* (PE) (see Fig. 2). A pair of memory units is used for each memory module. This double-buffering scheme allows data to be moved between one memory unit and secondary storage (the Memory Storage System) while the processor operates on data in the other memory unit.

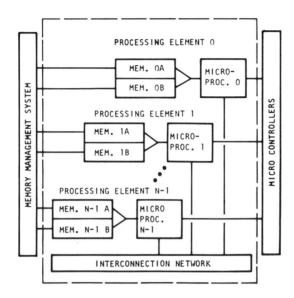

Fig. 2. PASM Parallel Computation Unit.

The *interconnection network* provides a means of communication among the PCU PEs. Two different interconnection networks are being considered for PASM: the generalized cube and the augmented data manipulator. Both consist of n stages of switches and are controlled by routing tags. Both can be partitioned into independent subnetworks if all of the PEs in a partition of size $P = 2^p$ have the same value in the low-order $n - p$ bit positions of their addresses. Studies are currently being conducted to choose which of these networks to implement in PASM.

The *Micro Controllers* (MCs) are a set of $Q = 2^q$ microprogrammable microprocessors, numbered (addressed) from 0 to $Q - 1$, which act as the control units for the PCU processors in SIMD mode and orchestrate the activities of the PCU processors in MIMD mode. Each MC is attached to a memory module (a pair of memory units so that memory loading and computations can be overlapped). *Control storage* contains the programs for the MCs.

Each MC controls N/Q PCU processors. The physical addresses of the N/Q

PEs connected to an MC have as their low-order q bits the physical address of the MC (see Fig. 3), so that the network can be partitioned. Possible values for N and Q are 1024 and 16, respectively. A virtual SIMD machine (partition) of size RN/Q, $R = 2^r$ and $0 \leq r \leq q$, is obtained by loading R MC memory modules with the same instructions simultaneously. In SIMD mode, the R MCs are synchronized and each MC fetches instructions from its memory module, executing the control flow instructions (e.g. branches) and broadcasting the data processing instructions to its PCU PEs. Similarly, a virtual MIMD machine of size RN/Q is obtained by combining the efforts of the PCU processors of R MCs. In both cases, the physical addresses of these MCs must have the same low-order $q - r$ bits so that all of the PCU PEs in the partition have the same low-order $q - r$ physical address bits.

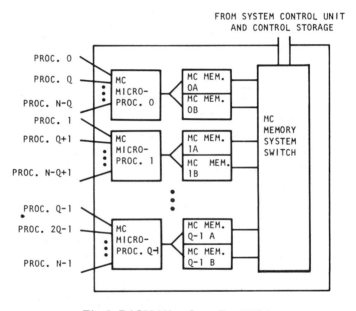

Fig. 3. PASM Micro Controllers (MCs).

In each partition, the PCU PEs are assigned *logical addresses*. Given a virtual machine of size RN/Q, the PEs have logical numbers 0 to $(RN/Q) - 1$ (the high-order $r + n - q$ bits of the physical number). Similarly, the MCs assigned are logically numbered from 0 to $R - 1$ (for $R > 1$, the high-order r bits of its physical number). The PASM language compilers and operating system will be used to convert from logical to physical addresses, so a system user will deal only with logical addresses.

When large SIMD jobs are executed, i.e. jobs which require more than N/Q

processors, more than one MC executes the same set of instructions. To load several MC memories with the same set of instructions at the same time, Control Storage is connected to all of the MC memory modules via a bus. Each memory module is either enabled or disabled for loading from Control Storage, depending on the contents of a Q bit register. Another Q bit register selects which memory unit an MC processor should use for instructions. An enabled memory unit not being used by an MC processor receives the data from Control Storage. Both of these registers are set by the SCU.

Instructions which examine the collective status of all of the PEs of a virtual SIMD machine include "if any", "if all" and "if none". These instructions change the flow of control of the program at execution time depending on whether or not any or all processors in the SIMD machine satisfy some condition. Specialized hardware to handle this task uses a bus which connects the MCs and requires that each MC have access to a $q + 1$ bit job *identification number* (ID) for the job it is executing. When an "if any" type instruction is encountered, each MC associated with the job sends a request to use the ID bus. It broadcasts its job ID to all of the MCs. If an MC is executing the job with the ID which is on the ID bus it then puts its local results onto the one bit data bus, which is constructed using "wired and" technology, i.e. the bus is a Q input "wired and" gate. This allows all of the MCs associated with a job to put their data on the bus simultaneously and then read the data and take the appropriate action.

The *Memory Management System* controls the loading and unloading of the PCU memory modules. It employs a set of co-operating dedicated micro-processors. The *Memory Storage System* provides secondary storage for these files. Multiple devices are used to allow parallel data transfers.

The Memory Storage System will consist of N/Q independent *Memory Storage units*, numbered from 0 to $(N/Q) - 1$. Each Memory Storage unit is connected to Q PCU memory units. For $0 \le i < N/Q$, Memory Storage unit i is connected to those memory modules whose physical addresses are of the form $(Q^{\star}i) + k, 0 \le k < Q$. Thus, Memory Storage unit i is connected to the ith processor/memory module pair of each MC (see Fig. 4). Since the PE memories are double-buffered, while one job is being processed, results from the previous job can be stored and the next may be loaded.

The two main advantages of this approach for a partition of size N/Q are that (i) all of the memory modules can be loaded in parallel and (ii) the data is directly available no matter which partition (MC group) is chosen. This is done by storing in Memory Storage unit i the data for a task which is to be loaded into the ith logical memory module of the virtual machine of size N/Q, $0 \le i < N/Q$. Thus, no matter which MC group of N/Q processors is chosen, the data from the ith Memory Storage unit can be loaded into the ith logical memory module of the virtual machine, for all $i, 0 \le i < N/Q$, simultaneously,

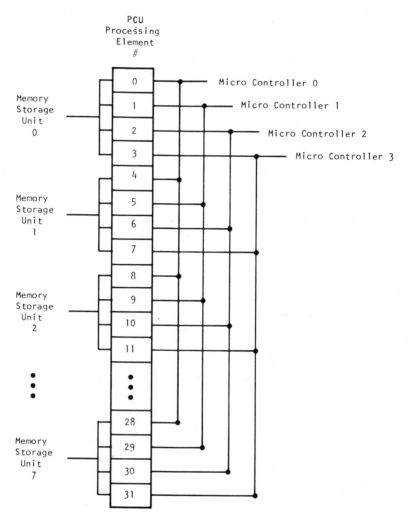

Fig. 4. Organization of the PASM Memory Storage System for N = *32 and* Q = *4, where "PCU" is Parallel Computation Unit.*

i.e. in one parallel block transfer. This same approach can be taken if only $(N/Q)/2^d$ distinct Memory Storage units are available, $0 \leq d < n - q$, using 2^d parallel block loads will be required instead of just one. In general, a task needing RN/Q processors, $1 \leq R \leq Q$, logically numbered 0 to $RN/Q - 1$, will require R parallel block loads if the data for the memory module whose high-order $n - q$ logical address bits equal i is loaded into Memory Storage unit i. This is true no matter which group of R MCs (which agree in their low-order $q - r$ address bits) is chosen. If only $(N/Q)/2^d$ distinct Memory Storage units are available, $0 \leq d < n - q$, then $R{\star}2^d$ parallel block loads will be required instead of just R.

There may be some cases where all of the data will not fit into the PCU memory space allocated. Assume a *memory frame* is the amount of space used in the PCU memory units for the storage of data from secondary storage for a particular task. There are tasks where many memory frames are to be processed by the same program. The double-buffered memory modules can be used so that as soon as the data in one memory unit is processed, the processor can switch to the other unit and continue executing the same program. When the processor is ready to switch memory units, it signals the Memory Management System that is has finished using the data in the memory unit to which it is currently connected. The Memory Management System can then unload the "processed" memory unit and then load it with the next memory frame.

To further increase the flexibility of PASM, a task may alter the sequence of data processed by it during execution. As an example, a task might examine a visible spectrum copy of an image and, based on features identified in the image, choose to examine an infrared spectrum copy of the same image. Rather than burden the System Control Unit to perform data loading sequence alterations, the task is allowed to communicate directly with the Memory Management System through the PE or MC.

A set of microprocessors is dedicated to performing the Memory Management System tasks in a distributed fashion, i.e. one processor handles Memory Storage System bus control, one handles the scheduling tasks, etc. This distributed processing approach is chosen in order to provide the Memory Management System with a large amount of processing power at low cost and high speed (due to the parallelism possible).

In summary, the organization of PASM, a large-scale partitionable SIMD/MIMD multimicroprocessor system for image processing and pattern recognition, has been briefly overviewed. For additional information about various aspects of PASM see: organization [9, 15, 16], instruction set [14], masking schemes for enabling and disabling PEs [7, 8, 15, 16], interconnection networks [4, 7, 10–12, 18, 20, 21], operating system [17], programming language [5], and memory management system [13, 16]. Examples of the use of parallel multimicroprocessor systems such as PASM are given in a companion paper [19]. In conclusion, a dynamically reconfigurable system such as PASM should be a valuable tool for both image processing/pattern recognition and parallel processing research.

REFERENCES

[1] Briggs, F., Fu, K. S., Hwang, K. and Patel, J. (1979). "PM4—a reconfigurable multimicroprocessor system for pattern recognition and image processing". AFIPS Nat. Comp. Conf. **48**, 255–265.

[2] Keng, J. and Fu, K. S. (1978). "A special computer architecture for image processing". Proc. Conf. Pattern Recog. and Image Processing, pp. 287–290.

[3] Lipovski, G. L. and Tripathi, A. (1977). "A reconfigurable varistructure array processor". Int. Conf. Parallel Processing, pp. 165–174.

[4] McMillen, R. J. and Siegel, H. J. (1980). "MIMD machine communication using the augmented data manipulator network". 7th Symp. Comp. Arch., pp. 51–58.

[5] Mueller Jr., P. T., Siegel, L. J. and Siegel, H. J. (1980). "A parallel language for image and speech processing". COMPSAC '80, pp. 476–483.

[6] Nutt, G. (1977). "Microprocessor implementation of a parallel processor". 4th Symp. Comp. Arch., pp. 147–152.

[7] Siegel, H. J. (1977). "Analysis techniques for SIMD machine interconnection networks and the effect of processor address marks". Trans. IEEE on Comp. **C-26**, 153–161.

[8] Siegel, H. J. (1977). "Controlling the active/inactive status of SIMD machine processors". Int. Conf. Parallel Processing, p. 183.

[9] Siegel, H. J. (1978). "Preliminary design of a versatile parallel image processing system". Third Biennial Conf. on Computing in Indiana, pp. 11–25.

[10] Siegel, H. J. (1979). "Interconnection networks for SIMD machines". *Computer* **12**, 57–65.

[11] Siegel, H. J. (1979). "A model of SIMD machines and a comparison of various interconnection networks". Trans. IEEE on Comp. **C-28**, 907–917.

[12] Siegel, H. J. (1980). "The theory underlying the partitioning of permutation networks". Trans. IEEE on Comp. **C-29**, 791–801.

[13] Siegel, H. J., Kemmerer, F. and Washburn, M. (1979). "Parallel memory system for a partitionable SIMD/MIMD machine". Int. Conf. Parallel Processing, pp. 212–221.

[14] Siegel, H. J. and Mueller Jr., P. T. (1978). "The organization and language design of microprocessors for an SIMD/MIMD system". 2nd Rocky Mt. Symp. on Microcomputers, pp. 311–340.

[15] Siegel, H. J., Mueller Jr., P. T. and Smalley Jr., H. E. (1978). "Control of a partitionable multimicroprocessor system". Int. Conf. Parallel Processing, pp. 9–17.

[16] Siegel, H. J., Siegel, L. J., Kemmerer, F., Mueller Jr., P. T., Smalley Jr., H. E. and Smith, S. D. (1979). "PASM—a partitionable multimicrocomputer SIMD/MIMD system for image processing and pattern recognition". Tech. Rep. TR-EE-79-40, School of Electr. Eng., Purdue University.

[17] Siegel, H. J., Siegel, L. J., McMillen, R. J., Mueller Jr., P. T. and Smith, S. D. (1979). "An SIMD/MIMD multimicroprocessor system for image processing and pattern recognition". Proc. Conf. Pattern Recog. and Image Processing, pp. 214–224.

[18] Siegel, H. J. and Smith, S. D. (1978). "Study of multistage SIMD interconnection networks". 5th Symp. Comp. Arch., pp. 223–229.

[19] Siegel, L. J. (1981). "Image processing on a partitionable SIMD machine". (This volume.)

Smith, S. D. and Siegel, H. J. (1978). "Recirculating, pipelined, and multistage SIMD interconnection networks". Int. Conf. Parallel Processing, pp. 206–214.

Smith, S. D., Siegel, H. J., McMillen, R. J. and Adams III, G. B. (1980). "Use of the augmented data manipulator multistage network for SIMD machines". Int. Conf. Parallel Processing, pp. 75–78.

Chapter Twenty-One

The Anatomy of VLSI Binary Array Processors

Anthony P. Reeves

1. INTRODUCTION

The main features of the Binary Array Processor (BAP) organization [1] result from the bit-serial architecture of the Processing Elements (PEs) and near-neighbour interconnection scheme. The bit-serial architecture allows flexible data formats and makes the BAP very efficient with respect to memory and processing resource utilization. Many image processing algorithms require that data within local areas of each pixel is to be combined; the near neighbour interconnection scheme enables these algorithms to be efficiently implemented.

A general block diagram for a large scale BAP system is shown in Fig. 1. The data processing is achieved by the array of Processing Elements which simultaneously process a whole image or a consecutive submatrix of an image. In the second case the total image is processed as a sequence of these submatrices. Data is input to and output from the PE array via the I/O buffer memory which communicates the data to image peripherals and conventional computer bulk storage devices. Instructions to the PE array are issued by a single high speed microprogrammed controller. The whole system synchronization is maintained by a conventional host computer which issues macro-instructions to the controller. Some feature information may be extracted from the PE array by the global information extraction mechanism.

With the advent of LSI technology the construction of highly parallel BAPs has become feasible. CLIP4 [2] is an operational 96 × 96 matrix parallel BAP. It involves a special NMOS chip which contains 8 PEs including 32 bits of local memory for each PE. The Massively Parallel Processor (MPP) is being built by Goodyear Aerospace for NASA [3]; it is scheduled to be operational by mid-1982. The 128 × 128 MPP PE array is constructed with special LSI

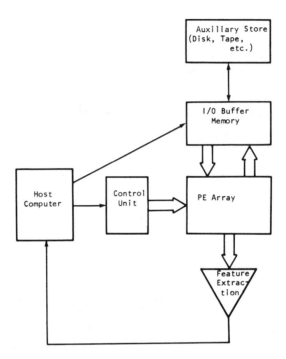

Fig. 1. Binary Array Processor System.

CMOS/SOS chips containing 8 PEs without local memory. A LSI version of the Distributed Array Processor (DAP) PE has been developed based upon the Uncommitted Logic Array (ULA) approach [4]. The chip contains 4 PEs without local memory and a 128 × 128 PE matrix processor is being considered.

2. CURRENT PE DESIGNS

The features of current PE designs will be discussed by describing two diverse current PE architectures. These are: the PE for the BASE system which is being developed in a small prototype form at Purdue University [1], and the PE for the MPP.

The BASE PE is shown in Fig. 2. It consists of three main components: a Boolean processor which can implement any three input Boolean function, a 1-bit wide local memory and a Near Neighbour Function processor (NNF). In general, the operands A, B and C would be held in 1-bit registers which could receive data from the local memory on a common bus line.

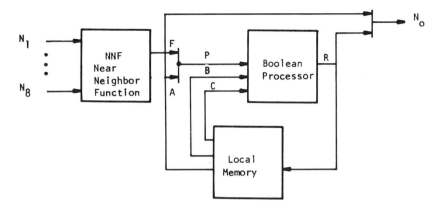

Fig. 2. The BASE PE organization.

The NNF receives data from the 8 near neighbour PEs and realizes the following function, which is similar to the CLIP4 near neighbour function:

$$F = \bigvee_{i=1}^{8} (g_i \wedge N_i)$$

where $\{g_1 \ldots g_8\}$ is an 8-bit control vector. A hexagonal near neighbourhood may also be selected.

There are three fundamental instruction types for BAPs [1]: Boolean, simple near neighbour and recursive near neighbour. Boolean instructions are used for logical and arithmetic operations within the local memory; the NNF is not used. The BASE Boolean processor can implement any of the 256 three input Boolean functions as specified by an 8-bit control vector. A 2-input Boolean function could be adequate for Boolean instructions; however, the 3-input function is much more efficient for arithmetic operations and is also necessary for recursive near neighbour instructions.

Arithmetic may be achieved with conventional bit-serial algorithms or, in some cases, functions may be more efficiently implemented with a specially optimized instruction sequence [5]. Simple near neighbour instructions specify that one operand comes from a near neighbour PE or a selected subset of near neighbours. An ORed subset of near neighbours is useful in some binary image algorithms.

In recursive near neighbour instructions the near neighbour output is taken from R rather than A in Fig. 2. A signal may propagate through a connected sequence of PEs in a single recursive instruction. This asynchronous operation may be several times faster than an equivalent sequence of simple near neighbour instructions depending upon the technology and detailed design of the processor. Recursive near neighbour instructions are useful for horizontal arithmetic, which treats the rows of the PE matrix as a set of data items, and some two dimensional binary topology operations.

A simplified MPP PE is shown in Fig. 3. The emphasis with this design is fast arithmetic computation rather than binary near neighbour operations. The NN select unit enables a bit-plane to be shifted in one of the four cardinal directions in one instruction. The Boolean processor implements all 16 possible Boolean functions between the P register and the value on the data bus; in this case P is an accumulator. For arithmetic operations a dynamically reconfigurable, variable length shift register with a maximum length of 30 bits and a full adder is available. This organization is faster than the BASE scheme especially where multiplication is involved; in this case the shift register is used for circulating the partial product. Multiplication still requires in the order of N^2 operations for N bit operands; however, the optimal number of memory cycles is 4 N, i.e. N memory cycles to read each operand and 2 N memory cycles to store the 2 N-bit result.

The S register is used for I/O and allows information to be shifted along the rows of the matrix independent of PE processing functions. The G register when set inhibits the processing functions of the PE. In this way instructions can be specified to be executed on a subset of PEs.

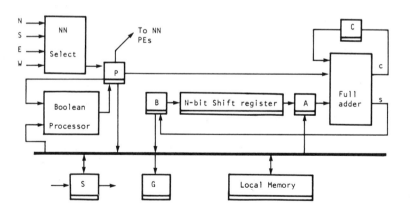

Fig. 3. Simplified MPP PE organization.

3. THE VLSI PE ORGANIZATION

The proposed VLSI PE array has three principal chip types as shown in Fig. 4. The first chip contains the PE ALU and some local memory (from 8–32 PEs on a chip is feasible). Data communication to each PE is by a single data line. The chip will have a very high complexity; it may be desirable to have one or two spare PEs on a chip to provide some fault tolerance. The interconnection network chip would select which subset of PEs to use. There are no data interconnections between PEs on the ALU chip, therefore they may be considered to be independent and may be assigned to arbitrary positions on the PE matrix.

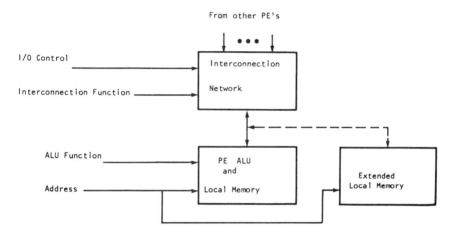

Fig. 4. VLSI PE organization.

The interconnection network chip contains the near neighbour function and an I/O mechanism. This chip realizes very simple switching functions and will be limited by the maximum number of possible pin connections. Several interconnection chips may be required for each ALU chip. The interconnection chip may be designed to have substantial fault tolerance due to the low function density.

The local memory on the ALU chip will be limited in size (from 1 K to 16 K bits per PE is feasible). For some system configurations it may be desirable to have a larger local memory. This may be easily achieved by adding conventional memory chips which are connected to the data links between the ALU chip and the interconnection chip.

A feasible VLSI ALU design is shown in Fig. 5. Logical, addition and multiplication operations can be achieved in an optimal number of memory

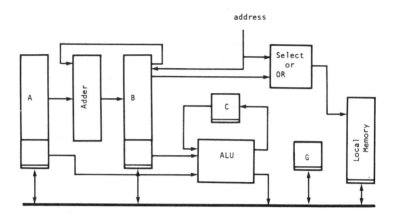

Fig. 5. VLSI PE ALU organization.

cycles. A and B are multi-bit registers to speed up the multiplication operation. The multiplicand is loaded into the A register and the partial products are accumulated in the B register by means of the parallel adder. This organization can achieve multiplication in the optimal 4 N clock cycles. The implementation of the parallel adder may involve problems due both to the delay and complexity of the parallel carry generation. The adder could feasibly be 32 or more bits long. A simple solution to this problem is to use a simpler carry-save adder for which complete carry generation is not necessary. The C register would have to be extended to store the intermediate carry signals. There are no delay problems with this scheme and the multiplication can still be achieved in the optimum 4 N clock cycles.

The second feature of the VLSI PE design is that the local memory is on the chip and may be addressed by the B register. This facility is not possible usually with simpler PE architectures as they do not involve the B register. Table-look-up algorithms may be efficiently implemented with this scheme. Moreover, conventional indexed addressing is possible by adding an index in A to a broadcast address in B. If a conventional parallel adder is implemented the indexed address calculation can be achieved in 1 clock cycle. However if the carry-save adder is implemented the index operation may require up to N clock cycles for an N-bit address.

When an external local memory is used it cannot be addressed with the B register; however, for many applications it is possible to arrange that all data to be indexed is stored in the internal local memory.

Since with the more-complex VLSI PEs multiple datapaths are possible, the choice of a bit-serial scheme should be reaffirmed. Assuming that the PE-memory bandwidth is a constant, one could have N bit-serial PEs or N/K

K-bit PEs at approximately the same cost. The number of data interconnection wires would also be the same in each case.

If the data to be processed is N or a multiple of N elements and the operations are optimal with respect to memory cycles, e.g. the proposed bit-serial multiplication scheme, then the arithmetic processing speed of the bit-serial scheme will be the same as with N/K K-bit PEs. There are several advantages, however, of implementing a bit-serial scheme instead of a K-bit fixed word format. With the bit-serial scheme the precision of the data may be varied by the user. This is useful when the data is not a multiple of K bits; the local store is more efficiently utilized and the processing speed is faster. If real-time speed constraints cause problems it is possible to trade computation precision with speed and memory. Dynamic variation of precision may also be possible. For example, if the difference of two similar arrays is computed the result may require less bits to represent than the operands. This will only be known at run time; a bit-serial scheme could be designed to only maintain the number of bits actually required. With the fixed format scheme data storage and processing resources are often inefficiently used.

Other advantages of the bit-serial scheme include direct mapping between grey-level and binary data and efficient implementation of unconventional data formats. Therefore, with the stated assumptions the bit-serial format offers many significant advantages over any K-bit scheme.

The exact design of the interconnection chip will depend upon the applications the system is to be used for. For shape analysis on binary images the BASE near neighbour functions are very effective; however, for grey-level data the MPP near neighbour functions are quite adequate. A rich set of multi-near neighbour functions can easily be implemented on the chip; unfortunately, many more pin connections are required. A general chip design is shown in Fig. 6—broken lines indicate optional features. I/O and fault tolerant features are not shown. The P register is not needed if there are both an input and an output connection to the PE ALU rather than a bidirectional bus. The Q register, the AND gate and the OR gate, are only required for recursive instructions. In this case a 1 in the P register initiates a signal and a 1 in the Q register allows a signal to propagate through the PE. The NN function selects one of the four near neighbours (or may compute a function of near neighbours). Other inputs to the NN function may come from other near neighbours if multi-near neighbour functions are implemented.

4. CONCLUSION

With the increased functional complexity of VLSI technology the following features may be found in future BAP PE designs: limited local memory, fault

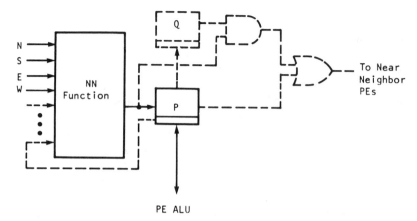

N →
S →
E →
W →
NN Function
Q
P
To Near Neighbor PEs

PE ALU

Fig. 6. General interconnection chip organization. Broken lines indicate optional sections.

tolerance, a limited table-look-up capability, and optimal execution speeds for arithmetic operations including multiplication. Furthermore, if the PE matrix is not larger than the dimensions of the image to be processed then bit-serial datapaths still offer the best utilization of resources without any loss of speed.

REFERENCES

[1] Reeves, A. P. (1980). "A systematically designed binary array processor". Trans. IEEE on Comp., **C-29,** 278–287.

[2] Duff, M. J. B. (1978). "Review of the CLIP image processing system". Nat. Comp. Conf., Anaheim, USA, pp.1055–1060.

[3] Batcher, K. E. (1980). "Design of a massively parallel processor". Trans. IEEE on Comp., **C-29,** 836–840.

[4] Reddaway, S. F. (1979). "The DAP approach". Infotech State of the Art Rep. on Supercomputers, **2,** 309–329.

[5] Reeves, A. P. and Rindfuss, R. (1979). "The BASE-8 binary array processor". Proc. Conf. on Patt. Recog. and Image Proc., Chicago, pp. 250–255.

[6] Reeves, A. P. and Bruner, J. D. (1980). "Efficient function implementation for bit—serial parallel processors". Trans. IEEE on Comp., **C-29,** 841–844.

Chapter Twenty-Two

The ICL DAP and its Application to Image Processing

D. J. Hunt

1. INTRODUCTION

The ICL Distributed Array Processor (DAP) began as a computer design for
large numerical computations of science and engineering. It has since been
recognized that its potential applicability is much wider; in particular, its even
higher power at non-numeric work makes it suitable for a variety of image
processing tasks.

Large problems commonly consume computer time by operating on arrays
of data. To match this structure, DAP has an array of identical Processing
Elements (PEs) each with local memory. They obey a common instruction
stream broadcast from a Master Control Unit (MCU). However, many
problems require some processing to be position dependent according to local
data conditions or a predetermined pattern. Hence, there is an "activity
control" effectively enabling individual PEs to be rapidly switched on and off.

The connectivity of the array is two-dimensional. This is both practical from
a construction point of view and corresponds directly to the most important
problem geometries. Other cases may be mapped onto such an array; for
example, a three-dimensional grid may be processed plane by plane, or for
smaller problems several planes dealt with simultaneously.

A fundamental decision in designing any such parallel machine concerns the
complexity of individual PEs. With DAP, the PEs are just one-bit wide as this
made the best use of a given amount of hardware by permitting a machine with
several thousand PEs. Thus, arithmetic operations are implemented by subroutines working bit-by-bit giving great flexibility of function and word length.
Performance is dependent on word length so short integers and booleans, both
common in image processing, can be operated upon particularly rapidly.

DAP is not a complete processing system, but attaches via a store highway to a conventional system acting as host. The PE memories together form a memory module of that host system, thus providing data communication with the host. Input–output is provided by the host together with program development functions such as editing and compilation.

The first DAP to be marketed has 4096 PEs arranged 64 × 64 and each PE has 4096 bits of memory making 2 Mbytes in total. Clearly, the principle is applicable to smaller and larger arrays. At present, standard MSI TTL circuitry is used for the PEs but such an architecture is ideal for LSI implementation.

2. PROCESSING ELEMENT ARRAY

Figure 1 shows a single PE together with its memory. This PE differs in detail from that of the "pilot" DAP previously published. All the data paths and registers are just one-bit wide. The only features omitted are certain enable inputs and details of the multiplexors.

The main component is a one-bit wide adder having three inputs:
(a) the Q register (or "accumulator");
(b) the C register (or "carry");
(c) the PE input (typically from local memory via the two multiplexors in succession).

In the general case, these inputs are added to produce a sum and a carry; the former may be written into Q and the latter into C. Thus, an integer add may be implemented by successively adding bits, intermediate carry's being held in the C register.

Other operations are possible. For example, loading Q from memory can be achieved by disabling the Q and C inputs to the adder. Also Q may be transferred to C by making the PE input True and disabling the C input to the adder. The upper multiplexor optionally inverts data from the lower multiplexor thus allowing subtraction, for example.

Register A implements the activity control as will be indicated later, and is also used for other purposes. The PE input may be written direct into A or AND-ed with the existing contents of that register.

The lower multiplexor in the figure selects the PE output. This may be the local memory, adder sum, Q register or A register. Conditional write to memory is achieved through a read-modify-write sequence. For this purpose A acts as a control input to the multiplexor rather than a data input, selecting either the adder sum or the old memory contents.

Each PE has a data connection to its four nearest neighbours in the array, making it possible to shift the entire matrix of Q registers in a northerly direction (say) by one place in a single machine cycle. To do this, it is arranged

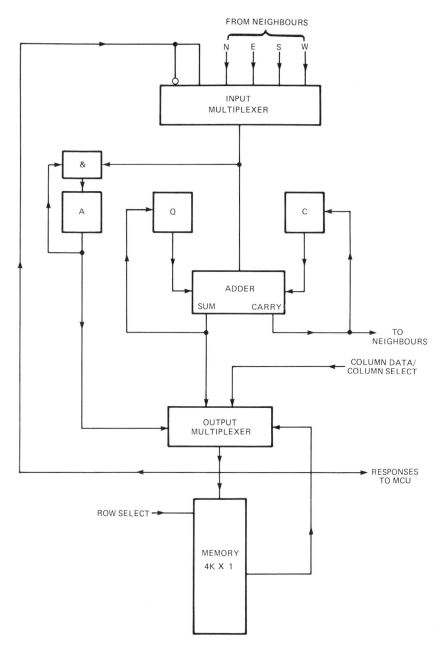

Fig. 1. PE diagram.

that the Q register contents appear at the carry output of the adder. In each destination PE the carry output from the appropriate neighbour is selected by the upper multiplexor and then passes through the adder again to the Q register.

It is sometimes appropriate for a row of PEs to cooperate in producing a single arithmetic result rather than the normal case of working independently. The neighbour connections already described also provide a carry path from PE to PE to facilitate this. One operand is placed in the set of Q registers, the other in the set of C registers (a change of role for C) and the carry's propagate asynchronously via the PE inputs in the specified direction. Multiple machine cycles are allowed for this, the carry propagating through at least four PEs in each cycle. Most commonly the carry's propagate from east to west in each row, but any direction is possible.

The PE memory is one-bit wide and is randomly addressable, but the address is common to all PEs. Normally data is held in a "vertical" format where successive bits of a given word are held in successive locations of a particular PE memory.

Data highways from MCU to array allow data to be broadcast across the entire array in one cycle in either of two orthogonal orientations. The column highway connects as a further input to the lower multiplexor. The row high-way connects to the enable inputs of the memory, but can also have the effect of a data input to the PE. A further form of broadcast sends a single bit from the MCU to every PE; this uses the PE control signals rather than the highways.

Data paths from array to MCU allow the AND function of PE outputs to be formed across all rows or all columns and input to the MCU in a single cycle. A single row (or column) may be returned to the MCU by sending a suitable pattern on row select (column highway). This causes the PE output to be True for all but the required row (column).

3. THE MASTER CONTROL UNIT

The MCU implements a stream of instructions by broadcasting control signals and addresses to the PEs. It also interfaces to the host system and diagnostic facilities.

Instructions are fetched from the array itself which for this purpose acts as a passive memory. A high proportion of instructions executed are in short loops needed for operating on successive bits of a word. Such loops are specified by a hardware instruction causing instructions within the loop to be buffered by the MCU. This dramatically reduces the instruction fetching overhead and allows automatic stepping of memory addresses without any time penalty.

The low-level programmer has access to eight registers within the MCU having length equal to the side of the array. Any register may be the source of data sent to the array along the row or column highways or the destination of the row or column AND functions. Also, the registers can hold an address in the form of a bit-plane number plus a row or column number. Facilities exist for integer add and logical functions on the register contents; these do not involve the array.

4. PROGRAMMING

Most programming employs an array processing extension of Fortran known as DAP-FORTRAN. This has matrices and vectors as basic elements as well as scalars. The language provides a good match between the capabilities of DAP and most requirements of applications and, at least for numeric work, the overheads in using this high-level language are small.

Boolean operations can also be expressed making DAP-FORTRAN suitable for development of image processing programs. However, for best performance on proven algorithms, an assembly language (APAL) is also available. The APAL programmer can improve performance by using the hardware assisted loop facility and by taking full advantage of the logical capabilities of the PE.

Although low-level programming needs detailed work, two factors increase its feasibility. First, DAP-FORTRAN and APAL can be mixed at the subroutine level, so that usually only critical routines need be converted. Secondly, APAL has comprehensive macro-facilities, enabling the user to define operations tailored to his requirements; in effect, he is setting up his own high-level language.

5. APPLICATION TECHNIQUES

DAP has the following features that make it suitable for many operations in image processing.
  (a) Bit-organized PEs are very versatile. There are large performance gains through use of short words and, particularly, for boolean work;
  (b) The two-dimensional structure is a direct match with most image processing requirements;
  (c) Highways allow rapid data movement;
  (d) The store is large enough to contain large images.
Use of DAP in some basic operations of image processing is illustrated in the following subsections; the list is not intended to be exhaustive.

5.1 Transforms

Each step of an FFT can be described algebraically as taking pairs of numbers
a and b and forming the new pair:

$$(a + \alpha b) \text{ and } (a - \alpha b)$$

where α is some predetermined constant.

For a two-dimensional transform of 64×64 data points, with the data
mapped in column major order onto the PE array, a pair of points a,b is
separated by the following distances in each of the 12 stages respectively:

	32, 16, 8, 4, 2, 1 PEs East–West
then	32, 16, 8, 4, 2, 1 PEs North–South

One step could be written as below in DAP-FORTRAN.

1. MASK = ALTC(K)
2. TEMP = SHWC (DATA,K)
3. TEMP (MASK) = DATA
4. DATA (MASK) = SHEC (DATA,K)
5. DATA = DATA + ALPHA * TEMP

Here DATA is a matrix containing the data being transformed, TEMP a
temporary matrix, ALPHA a matrix of coefficients as above, and MASK a
boolean matrix. Scalar K gives the amount of shift: 32, 16, 8, 4, 2 or 1.

Line 1 uses the function ALTC to set up MASK to be alternately False and
True in groups of K columns. Line 2 shifts the data west by K places and line 3
conditionally copies unshifted data using the PE activity control. Line 5
performs the arithmetic work, the operations + and * being applied on an
element-by-element basis to entire matrices. In fact, the whole step could be
expressed in a single statement using MERGE functions.

Advantage can be taken of the fact that multipliers α have a simple form for
the first two steps, and cyclic shifts of 32 in opposite directions are equivalent.
The code can easily be extended to improve performance through use of radix 4
or radix 8 techniques. Also larger problems can be dealt with by having several
data points per PE. These perform relatively better since it is arranged that
each FFT "butterfly" is entirely within a given PE.

A 4096 point real complex FFT programmed in DAP-FORTRAN takes
about 21 ms. Other transforms can be very much faster. For example,
Hadamard transforms involve no multiply and are particularly fast when
applied to short integers. DAP has the flexibility to be applied to Fermat
Number Transforms using special low-level routines for modulo arithmetic.
Such transforms may be applied to certain types of convolution.

5.2 Local operations

Applications involving window operators make good use of the PE nearest neighbour connections. Although the hardware only provides access to four neighbours, there is little difficulty or time penalty in accessing the diagnoal elements on a 3 × 3 window or extending to a 5 × 5 window. Hexagonal grids may also be used.

A trivial example is that of evaluating the AND function of all nine values on a 3 × 3 window. This can be achieved for a 64 × 64 grid as follows:

1. TEMP = INPUT .AND. SHEP (INPUT) .AND. SHWP (INPUT)
2. OUTPUT = TEMP .AND. SHNP (TEMP) .AND. SHSP (TEMP)

Here, INPUT, TEMP and OUTPUT are logical matrices. The shift functions move the operand in the specified direction, implicitly by one place only. In this case, the shifts are specified as planar (suffix "P" on the function name) meaning that zero (or False) is shifted in at the edges. Note, that in this particular case, advantage is taken of the fact that each intermediate value (in TEMP) contributes to three result values. Execution time would be about 17 μs in DAP-FORTRAN, or under 5 μs in APAL.

5.3 Re-mapping

Image re-mapping to correct for distortion or rotation may be implemented using one of two techniques.

When the distortion is large, the re-mapping can be considered as a table look-up. Each point of the input image in turn is extracted and re-broadcast to all PEs. Each PE holds a boolean mask determining whether it needs that data item and uses it to conditionally write the data to the output image. Of course, if certain areas of the input image are not needed at all, then work on them can be omitted.

If the distortion is small in the sense that each input point is near its destination then a scheme involving shifting the entire matrix around can be used. After each step of shift, data that is in the correct PE is copied to the output, and shifting continued, perhaps in a spiral pattern, until the complete output image has been formed. A generalization is to perform gross shifts on sections of the image until each data point is close to its destination before performing the spiral shift on the entire matrix.

Each method requires boolean matrices to control writing data. These are derived as needed from integers held in each PE specifying the address of data in the input image.

Interpolation may be applied to each method. This involves each PE receiving a few adjacent data items rather than just one. It is simple to arrange that in the first method each data item is only broadcast once, and in the second there is only one conditional copy after each stage of shift.

5.4 Global properties

Operations that evaluate global properties of a matrix are often faster than might be expected. For example, to find the maximum element of a real or integer matrix takes 60 μs. Using APAL coding, the method proceeds as follows for unsigned integers; extension to signed and real numbers is trivial. Initially, register A is True for those elements that are to be considered in the MAX function, often the entire matrix. The algorithm examines the matrix of operands bit-by-bit beginning with the matrix of most significant bits.

Each step begins by copying the current contents of register A into register Q. Then the matrix of operand bits at the current bit position is AND-ed into A. If now the OR function of all the A registers is True then, at least one of the operand values still under consideration has that bit set. Hence, the corresponding result bit is set and register A remains unchanged as a mask for the next step. Otherwise, no operand has that bit set so the result bit is unset and the previous mask is restored by copying Q into A.

Summation of a matrix is achieved by adding numbers in pairs, then the results in pairs and so on, there being 12 such steps on a 64 × 64 DAP. However, this can be very much faster than 12 general additions. First, for floating point operands, all input values can be initially normalized to a common exponent and all addition operations treated as fixed point although a few more guard bits than usual are used. Secondly, as the additions proceed, a number of PEs can cooperate in producing each intermediate result. The overall effect is that a floating point element-by-element addition of two matrices takes 180 μs, whereas summation of a single matrix takes 450 μs.

Performance can be better in relation to the number of input values when either all elements of a set of matrices are to be summed or when several matrices are to be summed independently. A special case of the latter is in histogram evaluation where a set of logical matrices is used, each sum being one element of the result.

6. CONCLUSIONS

DAP is very versatile. It's applications range from numerical work to bit manipulation and it is well suited to the wide range of operations in image processing.

Chapter Twenty-Three

CLIP4: A Progress Report

T. J. Fountain

1. INTRODUCTION

The detailed design of the CLIP4 image processing computer, shown in Fig. 1, has been widely described elsewhere [1], [2] and [3]. An outline of the important features will be given here, before describing the current status of the machine and attempting to point out some generally applicable conclusions arrived at during its manufacture.

2. THE CLIP4 SYSTEM

The central element of the machine is a two-dimensional array of simple Boolean processors, each communicating with its neighbours, each processor performing the same sequence of instructions simultaneously. The array is surrounded by circuits whose functions are to program the array, and to handle data input/output functions. The original data source for the array will usually have been a two-dimensional scene of some kind (e.g. microscope slide, Landsat image, etc), often input via a TV camera. The machine may accordingly be described at three (nested) levels.

At the first level, the processor cell circuit is shown in Fig. 2A, B, C and D are local data storage, A being used for data I/O. The two separately programmable Boolean processors each produce a function of A and P inputs, processor 1 giving the result of a cell operation, processor 2 giving a propagation function which is passed to cell neighbours. Each cell receives six or eight neighbour inputs, combinations of which are selected by the input gating to produce function T. Combinational logic produces a composite output P from inputs T, B and C.

At the second level, 9216 such cells are connected into a 96 × 96 point array

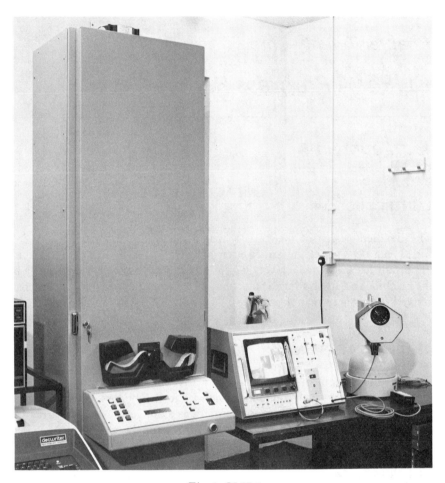

Fig. 1. CLIP4.

in either square mode (Fig. 3a) or hexagonal mode (Fig. 3b). In addition to the
neighbour connections shown in Fig. 3, a control bus passes to each processor
in the array.

The CLIP4 array is implemented in custom-designed LSI circuits, each of
which comprises eight processors and some local control logic. This technique
of implementation was, and still is, regarded as necessary for the manufacture
of a machine of this type, but this aspect of the development has provided the

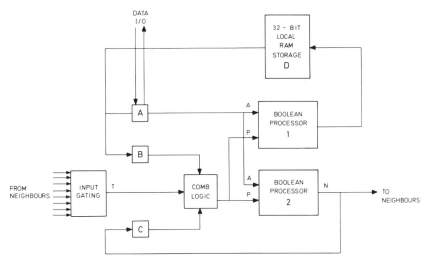

Fig. 2. Processor cell.

bulk of the problems encountered in this program. These problems will be described in some detail in section 4.

At the third level, the array is embedded in data interface and control circuits as shown in Fig. 4. A frame of grey-level data, usually taken from a TV camera, is frozen and stored externally to the array, and binary subsets of this data can be extracted and passed to the processor array. The array output is also stored externally and is available for display in a number of formats.

3. THE CURRENT STATUS OF CLIP4

CLIP4 is now working and is programmable under systems software running on a PDP11/10. With the exception of a small and decreasing number of control bugs, the only problems with the system are concerned with the performance of the custom circuits, considered more fully in section 4. An increasing number of application programs are being successfully developed on CLIP4, covering amongst others the following areas:

Grey-level picture segmentation

Analysis of moon crater size-frequency distribution

Analysis of amoeboid motion

Recombination of binocular image pairs

Moving scene analysis.

An example of grey-level segmentation as applied to a photograph of the lunar surface is shown in Fig. 5. The original is shown at 5(a). The analysis of

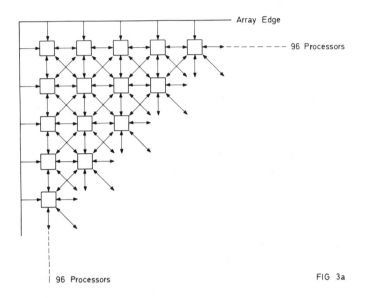

Array Edge

96 Processors

96 Processors

FIG 3a

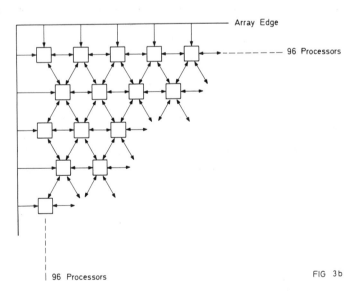

Array Edge

96 Processors

96 Processors

FIG 3b

Fig. 3. (a) Square connectivity; (b) hexagonal connectivity.

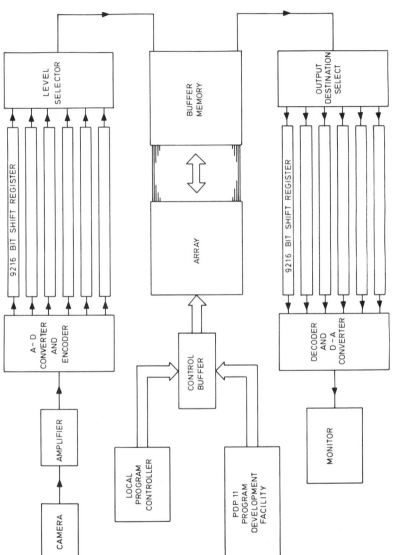

Fig. 4. CLIP4 system.

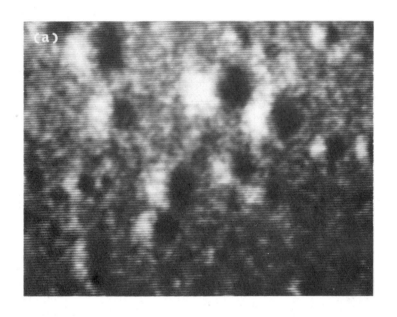

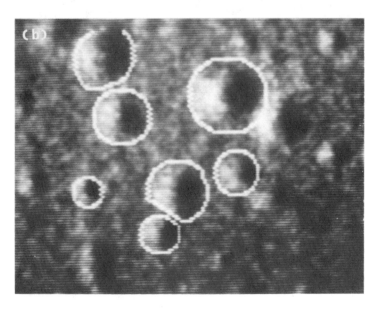

Fig. 5. Grey-level segmentation. (a) Input picture; (b) result.

the picture proceeds in a series of cycles, each cycle searching for, and marking with a circle, craters of a given diameter. Fig. 5(b) shows the pictorial result of the complete analysis. The processed area does not cover the whole picture, and craters which intersect with the edges of the array are ignored by the program.

The interactive software package, CRISP, is proving invaluable for program development and the ability of CLIP4 to perform large amounts of data processing at very high speeds leads to a very satisfactory user interaction with developing programs.

4. THE CLIP4 CHIP

The story of the development of the CLIP4 integrated circuit is a salutary one and may provide some valuable pointers for those contemplating the custom I.C. route to a final product. The approximate timetable was as follows:

Year	Month	
1973	Oct.	Discussions commenced with design house A.
1975	Jan.	Science Research Council grant announced. Design house A begins design.
1976	Dec.	Design house A produces a non-viable design.
1977	June	Contract placed with design house B.
1978	June	First prototype device from design house B with some faults.
1978	July	Design house B unable to continue work.
1978	Sept.	Contract placed with design house C to complete the design begun by B.
1979	May	Second prototype produced and testing commenced.
1979	July	Correct logical behaviour verified.
1980	Jan.	1400 chips delivered to a reduced specification.
1980	Feb.	Array of chips (1152) assembled.
1980	Oct.	1056 working chips remain in reduced array.

It is instructive to compare this seven-year development with that widely quoted in many parts of the semiconductor industry of one year from design to product. We feel that we have learned a number of useful facts during this program:

(1) Modern (fast, dense, etc) semiconductor technologies are not available on the custom design path unless sales of 100,000+ units per year are guaranteed.

(2) The most modern technologies are never available on the custom design path.
(3) With a circuit of worthwhile complexity, it is almost certain that at least one redesign cycle will be required to attain a logically correct circuit.
(4) Yields of large-scale integrated circuits are problematical and are a complex function of the required performance.
(5) It is almost impossible to design a complete test program for a complex circuit before "hands on" experience of the circuit has been gained. There may also be problems of running a test program in economical time.
(6) Burn-in of circuits is an important, time-consuming and frustrating process, and may leave one with insufficient parts.
(7) Characterization of circuits is a statistical process, dependent on a large number of circuits being produced, and is hence hardly a valid concept for custom design.

It is perfectly possible that, as in the case of CLIP4, custom design is the only feasible way of proceeding (even with the advantage of hindsight) and all the above factors have to be accepted. However, forewarned is forearmed, and a knowledge of the possible pitfalls may help one to plan a path through the minefields.

The current status of the CLIP4 chip can be summarized as follows:
(a) The logic of the chip functions correctly.
(b) 1056 working chips are available, compared with 1152 required for a full array.
(c) The clock rate is 1 MHz compared with 2.5 MHz specification.
(d) The chip design is uncharacterized with respect to temperature and voltage variations.
(e) A further batch of chips is being procured to act as spares and to obtain a full size array.
(f) An alternative processing source is being evaluated.
(g) The present array of chips is functioning reliably.

5. SUMMARY

The CLIP4 system is the largest working processor array yet constructed and is performing usefully on a variety of image processing problems. Difficulties involved in the procurement of custom integrated circuits have been overcome and the small number of circuits required to complete the full array area should soon be available.

REFERENCES

] Duff, M. J. B., Watson, D. M., Fountain, T. J. and Shaw, G. K. (1973). "A cellular logic array for image processing". *Patt. Recog.* **5**, 229–247.

] Duff, M. J. B. (1979). "Parallel processors for digital image processing". In *Advances in digital image processing*, Stucki, P. (ed.). Plenum Pub. Corp., New York, pp. 265–276.

] Fountain, T. J. and Goetcherian, V. (1980). "CLIP4 parallel processing system". IEE Proc., **127**, Pt.E, 219–224.

Chapter Twenty-Four

Image Processing on a Partitionable SIMD Machine*

L. J. Siegel

In Chapter 20, PASM, a partitionable SIMD/MIMD system being designed for image processing and pattern recognition applications, was described. In order to determine the architectural requirements for PASM, a variety of parallel algorithms to perform image processing tasks have been developed and analysed [1–3, 6, 7, 10]. The algorithms can be used to define features such as the number of processors needed/useful for a class of problems, the sizes of memories required, interconnection network capabilities needed, and the type of processing capability required in each processor. In this chapter, three SIMD image processing algorithms are presented, and the architectural requirements which can be inferred from the algorithms are investigated.

For the purpose of describing the algorithms, a skeletal model of an SIMD machine will be defined, and the algorithms will be used to add details to the model. The basic SIMD machine will consist of a control unit, N processing elements (PEs) each, a processor with its own memory, and an interconnection network. The PEs are addressed (numbered) from 0 to $N - 1$. The control unit broadcasts instructions to all PEs, and each active PE executes these instructions on the data in its own memory. Each instruction is executed simultaneously in all active PEs. The interconnection network allows interprocessor communication.

* This research was supported by the Defense Mapping Agency, monitored by the United States Air Force Rome Air Development Center Information Sciences Division, under Contract No. F30602-78-C-0025 through the University of Michigan.

1. SMOOTHING

In image smoothing, each pixel is assigned a grey-level equal to the average of the grey-levels of the pixel and its eight nearest neighbours in the unsmoothed image. This operation is performed for each pixel in the image, with the possible exception of the edge pixels. Smoothing of an $M \times M$ image will therefore require approximately M^2 smoothing operations, where each smoothing operation typically consists of eight additions and one division. In an SIMD algorithm using N PEs [6], the $N = r^2$ PEs are logically configured as an $r \times r$ grid, on which the $M \times M$ image is superimposed, i.e. each processor holds a $M/r \times M/r$ subimage. This is shown in Fig. 1(a). Smoothing is performed on each subimage in parallel, with data transfers needed to smooth the boundary points in each subimage. The number of smoothing operation steps performed will be $(M/r)^2 = M^2/N$. The number of parallel data transfer steps required will be $(4 M/r + 4)$: each edge of the subimage requires M/r pixels from the adjacent PE, and each corner pixel requires one pixel from a "diagonal" PE. This is shown in Fig. 1(b).

From the smoothing algorithm and application dependent performance requirements, some architecture characteristics of an SIMD machine to perform smoothing efficiently can be inferred. For an $N - PE$ machine, $N = r^2$, the SIMD algorithm will require M^2/N smoothing operations and $4 M/r + 4$ data • transfer operations. Given typical image sizes and required processing rates (e.g. images per minute), acceptable values for N can be selected. The PE processors need only to be able to perform integer arithmetic. Each PE memory will be required to hold $2 M^2/N + 4 M/r + 4$ pixels: the original and

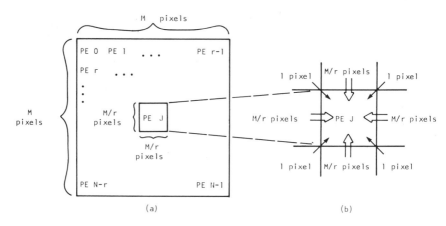

Fig. 1. (a) Data allocation for smoothing an M $\times$ M *image using* N $=$ r² *PEs. (b) Data transfers needed to smooth edge pixels in each* PE.

smoothed subimages, plus the border pixels transferred from adjacent PEs. The interconnection network will be required to perform transfers to a PE's eight nearest neighbours, when the PEs are viewed as being configured in the $r \times r$ grid. Since several (M/r) pixels are transferred from a PE to one of its neighbours, the ability to perform "block transfers" can decrease the time required for transfers. Masking is needed in order to disable the PEs at the edges of the $r \times r$ grid (e.g. to disable the right edge from performing the PE j to PE $j + 1$ transfers). The masking is based on the addresses of the PEs (e.g. all PEs whose address (number) is $(r - 1)$ mod r should be disabled for the PE j to PE $j + 1$ transfers). Each PE must therefore know its own address. Masking schemes capable of disabling the correct sets of PEs include PE address masks [4] and data conditional masks [8].

2. HISTOGRAM

Consider computing the B-bin histogram of an $M \times M$ image using $N \geq B$ PEs, $N = r^2$ [6]. Each pixel of the image is represented by a grey-level integer value between 0 and $B - 1$. The final histogram contains value j in bin i if exactly j of the pixels have grey-level i, $0 \leq i < B$.

As in the smoothing algorithm, the PEs are logically configured as an $r \times r$ grid, with each PE holding a $M/r \times M/r$ subimage. The "local" histograms are computed in parallel, so that each PE contains a B-bin histogram for the subimage which it holds. These local histograms are then combined using a form of overlapped recursive doubling. This is shown for $N = 16$ and $B = 4$ in Fig. 2.

In the first $b = \log_2 B$ steps, each block of B PEs performs B simultaneous recursive doublings [9] to compute the histogram for the portion of the image contained in the block. At the end of the b steps, each PE has one bin of this partial histogram. This is accomplished by first dividing the B PEs of a block into two groups. Each group accumulates the sums for half of the bins, and sends the bins it is not accumulating to the group which is accumulating those bins. At each step of the algorithm, each group of PEs is divided in half such that the PEs with the lower addresses form one group, and the PEs with the higher addresses form another. The accumulated sums are similarly divided in half based on their indices in the histogram. The groups then exchange sums, so that each PE contains only sum terms which it is accumulating. The newly received sums are added to the appropriate sums already in the PE. After b steps, each PE has the total value for one bin from the portion of the image contained in the B PEs in its block.

The next $\log_2(N/B)$ steps combine the results for these blocks to yield the histogram of the entire image spread out over B PEs, with the sum for bin i in processor i, $0 \leq i < B$. This is done by performing $\log_2(N/B)$ steps of a

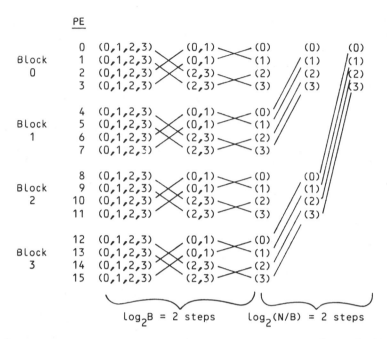

Fig. 2. Histogram calculation for N = 16 PEs, B = 4 bins. (w, . . .,z) denotes that bins w through z of the partial histogram are in the PE.

recursive doubling algorithm to sum the partial histograms from the N/B blocks. This is shown by the last two steps of Fig. 2.

A sequential algorithm to compute the histogram of an $M \times M$ image requires M^2 additions. The SIMD algorithm uses M^2/N additions for each PE to compute its local histogram and $B - 1 + \log_2(N/B)$ steps (transfer and add) to merge the histograms into the first B PEs. At step i in the merging of the partial histograms, $0 \leq i < \log_2 B$, the number of data transfers and additions required is $B/2^{i+1}$. A total of $B - 1$ transfers are therefore performed in the first $\log_2 B$ steps of the algorithm. $\log_2(N/B)$ parallel transfers and additions are needed to combine the block histograms. This technique therefore requires $B - 1 + \log_2(N/B)$ parallel transfer/add operations, plus the M^2/N additions needed to compute the local PE histograms.

An SIMD machine to execute the histogram algorithm requires $N \geq B$ PEs, N a power of two, $N = r^2$. For $M^2 \gg B$, the execution time will be proportional to $M^2/N + \log_2 N$. The PE processors need only perform integer addition. Each PE memory must hold M^2/N subimage pixels plus the B-bin local histogram. The interconnection network must perform the exchange-type transfers needed for the overlapped recursive doubling.

3. TWO-DIMENSIONAL FFT

As an example of the applicability of parallel computations to a different type of image processing task, the two-dimensional Fast Fourier Transform (FFT) is considered [3, 7]. The two-dimensional Discrete Fourier Transform (DFT) of an $L \times M$ array of elements $S(l,m)$ is commonly obtained by computing L one-dimensional M-point transforms on the L rows of the S array, to produce an intermediate array G, then computing M one-dimensional L-point transforms on the M columns of the G array. Figure 3 outlines an efficient SIMD method for performing the DFT on an $M \times M$ array S using M PEs. It is assumed that initially PE i contains row i of the array S. The DFTs on the rows of S are performed by executing a serial FFT in each PE, on the row of S contained in that PE. This serial FFT can be executed simultaneously by each of the M PEs. The resulting G array has row i in PE i. A transpose operation is performed on G, to rearrange the array so that PE i contains column i of G. A serial FFT executed in parallel in each PE performs the DFT on the columns of G, producing the (transpose of the) transform array F.

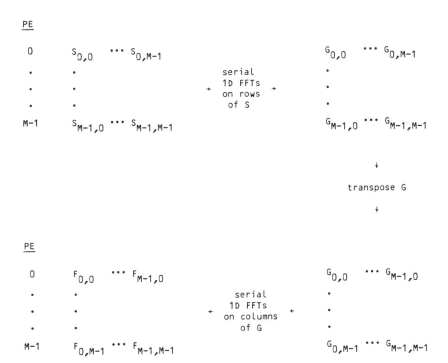

Fig. 3. Computation of two-dimensional FFT of M × M *array* S *using* M *PEs.*

298   *L. J. Siegel*

The basic operation in performing the transpose is the transfer of array element $G(j,k)$ from PE j to PE k. This is done for M $G(j,k)$'s in parallel, using an interconnection function which sends data from PE j to PE $(j + i)$ mod M for all of the $G(j,k)$ for which $(k - j)$ mod M is equal to i. This is shown for $M = 4$ and $i = 1$ in Fig. 4. The parallel transfer operation is performed for $1 \leqslant i < M$. (For $i = 0$, no transfer is needed; i.e., the diagonal of the transpose matrix is the same as the diagonal of the original matrix.) For each i value, the element which PE j sends is the kth element of the row of G held in PE j, where $k = (j + i)$ mod M. That element, received in PE k, is stored as the jth element of the column of G transpose being created in PE k, where $j = (k - i)$ mod M.

A serial algorithm to compute the DFT of an $M \times M$ array requires $M^2\log_2M$ complex multiplications. The parallel two-dimensional FFT algorithm presented requires $M\log_2M$ parallel complex multiplications and $M - 1$ parallel data transfers. The algorithm can be generalized to use $N = M/2^k$ PEs, $0 \leqslant k < \log_2M$. In this case, the number of complex multiplication steps is $2^kM\log_2M$, and the number of transfer steps is $2^kM - 2^{2k}$. The machine size N can therefore be chosen to meet performance requirements. The PE processors are required to perform complex arithmetic. To perform the transpose, each PE must have an independent indexing capability in order to select the element to transfer, and to be able to store correctly the element it receives. For example, assume each PE holds a vector GV containing its row of G, and a vector G^TV for its row of G transpose. Then for the $+1$ mod M transfer, PE 0 must retrieve $G(0,1)$ from position 1 of its GV vector; after the transfer, PE 1 must store $G(0,1)$ in position 0 of its G^TV vector. At the same time, PE 1 accesses $G(1,2)$ from position 2 its GV vector, and after the transfer, PE 2

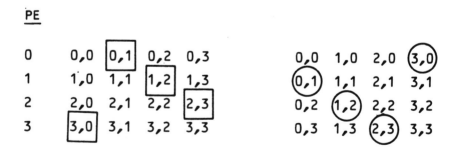

Fig. 4. Example of data transfers performed to transpose array G. Boxes indicate G(j,k)'s for which (k − j) mod M = 1. These elements are transferred simultaneously using a + 1 mod M transfer, and are stored in the circled locations in G transpose.

stores $G(1,2)$ in position 1 of its G^TV vector (see Fig. 4). Each PE must therefore be able to perform operations such as indexing based on knowledge of its PE number. Each PE memory must hold at most 4 $M2^k$ values (the image row, G row, G transpose row, and transform column), plus M constants needed to compute the one-dimensional FFTs. To transpose the G array, the interconnection network must perform "uniform shifts" of data; i.e. each PE j must be able to send data to PE $(j + d)$ mod N. Such transfers will be needed for $1 \leqslant d < N$.

CONCLUSION

Three SIMD algorithms to perform image processing operations have been presented and analysed. These and similar studies are being used to determine the architectural requirements for PASM, an SIMD/MIMD system for image processing and pattern recognition applications. Guided by the algorithm analyses, design choices can be made in order to provide the capabilities needed for the image processing/pattern recognition problem domain, while avoiding the costs and complexities of a general purpose system.

REFERENCES

] Feather, A. E. and Siegel, L. J. (1980). "SIMD algorithms to perform two-dimensional image correlation". Proc. 18th Allerton Conf. Communication, Control and Computing, University of Illinois, p. 985.

] Feather, A. E., Siegel, L. J. and Siegel, H. J. (1980). "Image correlation using parallel processing". 5th ICPR, Miami Beach, USA, pp. 503–507.

] Mueller Jr., P. T., Siegel, L. J. and Siegel, H. J. (1980). "Parallel algorithms for the two-dimensional FFT". 5th ICPR, Miami Beach, USA, pp. 497–502.

] Siegel, H. J. (1977). "Analysis techniques for SIMD machine interconnection networks and the effects of processor address masks". IEEE Trans. on Comp., C-26, 153–161.

] Siegel, H. J. (1981). "PASM: A reconfigurable multimicrocomputer system for image processing". (This volume.)

] Siegel, H. J, Siegel, L. J., McMillen, R. J., Mueller Jr., P. T. and Smith, S. D. (1979). "An SIMD/MIMD multimicroprocessor system for image processing and pattern recognition". Proc. Conf. Pattern Recognition and Image Processing, Chicago, USA, pp. 214–224.

] Siegel, L. J., Mueller Jr., P. T. and Siegel, H. J. (1979). FFT algorithms for SIMD machines". Proc 17th Allerton Conf. Communication, Control and Computing, University of Illinois, pp. 1006–1015.

] Siegel, H. J., Siegel, L. J., Kemmerer, F., Mueller Jr., P. T., Smalley Jr., H. E. and Smith, S. D. (1979). "PASM: A partitionable multimicrocomputer SIMD/MIMD system for image processing and pattern recognition". Int. Rep. TR-EE-79-40, School of Elect. Eng., Purdue University.

[9] Stone, H. S. (1975). "Parallel computers". In *Introduction to computer architecture*, Stone, H. S. (ed.). SRA, Chi., pp. 318–374.

[10] Swain, P. H., Siegel, H. J. and El-Achkar, J. (1980). "Multiprocessor implementation of image pattern recognition: a general approach". 5th ICPR, Miami Beach, USA, pp. 309–317.

Chapter Twenty-Five

Processing Capabilities Needed in Learning Systems for Picture Recognition

E. Persoon

1. LEARNING PICTURE RECOGNITION SYSTEMS

In picture recognition it becomes more and more a necessity that recognition systems have the ability to *learn* to recognize. In other words they should be able, when presented with one or more instances of objects that have to be recognized later, to extract the necessary information and form models of those objects. Using those models recognition is possible at a later moment.

The necessity for such systems can be found in the industrial process of the near future where picture recognition can aid in three major areas: automated product assembly, automated inspection and automated product handling. A recognition system to be used in those areas will have to be very flexible since it must be able to handle a wide variety of different industrial objects. Moreover it should be easily changeable from one recognition task to another.

It is obvious that such a system can be provided only with the basic algorithms used for learning and recognition and that detailed model knowledge about the objects to be recognized must be learned by the system itself. Following this line of thinking, a system has been conceived that learns to analyse an image in two steps. In the first step elementary image parts are learned and in the second step they are combined to learn complete objects. For detailed information on the first step refer to [1] and information on the second step can be found in [2]. The elementary image parts have a size of 11×11 pixels and are used as templates for analysing further images. So far template matching has not been very popular in picture recognition except for some specific applications. However the classical template matching can be made much more powerful by using the appropriate "similarity measure" for comparing the template with an input image and also by using the right size of template [1, 2].

If one agrees that some kind of "template matching" will be used in a learning system, then the speed with which such a system will be able to analyse an image will mainly depend on how fast a "template matching" can be performed. When implemented on standard computers, the recognition times of such a system are too long and it is without any doubt that special processors must be built and used to achieve useful recognition times.

Some of the requirements that such a processor should meet are presented in the next section.

2. SPECIAL PURPOSE PROCESSORS

In current processors a 3×3 neighbourhood is usually chosen for fast binary feature matching. However this is probably too small a neighbourhood. In some processors a 12×12 pixel neighbourhood is chosen which seems more suitable [3]. We feel it is important to have a processor on which different template matching algorithms can be implemented for a template size up to 12×12 pixels. The processor should not only work with binary images but also with grey-level images with up to 7 bits per pixel. Although many objects can be imaged in binary images several recognition tasks in the industrial environment require grey-level images.

The processor should be able to recognize templates at different positions in the image. So far this has been achieved by using the format of the image source (e.g. TV camera) as the main constraint. Using line buffers (to buffer several image lines) and parallel matching hardware, sufficient speed has been achieved [3]. However such an approach has many disadvantages. It is important not to be limited by the scan formats of the image source. When this is the case one can move the template over the input image in any desired sequence (for example, as needed in boundary tracing). To achieve this an image memory is needed that delivers very fast all pixel information inside a 12×12 square, positioned anywhere in the input image to be analysed.

Besides being able to perform template matching at different positions, the processor should be able to look at the image data at different resolutions. This is often necessary to attain optimal noise removal. To achieve this filtering is often needed to avoid aliasing effects. Therefore it is desirable that the processor can access easily larger-sized images (e.g. 24×24 pixels) and reduce them to smaller sized ones (12×12 pixels) by filtering and subsampling.

Another requirement is the capability to work with grey-level images. Frequently it is possible to derive from such an image a binary image by dynamic thresholding. The processor should be able to perform this operation quickly.

So far some of the capabilities of the processor have been mentioned that have to do with speed performance. In effect all the above operations can be done on a standard minicomputer but at insufficient speed.

Another aspect of the processor is its programming. Without any doubt, the processor will have to use parallellism to achieve the required speed. However, programming such a parallel computer is usually difficult. It is our opinion that such a processor can only be successful if it can easily be programmed. The programming language should hide from the user the irrelevant hardware structure of the processor.

3. SUMMARY

Some requirements for a processor to be used in a learning picture recognition system have been discussed. Such a processor is necessary to be able to introduce advanced pattern recognition techniques in the industrial environment.

REFERENCES

1] Persoon, E. (1978). "Principles for self organization and their application to picture recognition". Proc. 4th IJCPR, Kyoto, Japan, pp. 379–383.

2] Persoon, E. (1978–79). "A system that can learn to recognize two-dimensional shapes". *Philips Tech. Rev.* **30,** No. 11/12, 356–363.

3] Kashioka, S. and Ejiri, M. (1979). "A transistor wire-bonding system utilizing multiple local pattern matching techniques". Trans. IEEE on Syst., Man. and Cyb. **SMC-6,** pp. 562–570.

Chapter Twenty-Six

Comparison of Parallel Processing Machines: A Proposal

Kendall Preston, Jr

1. INTRODUCTION

Comparing different parallel processing machines is a difficult task. Machines may be compared according to their architecture: (i) SISDS (Single Instruction Single Data Stream); (ii) SIMDS (Single Instruction Multiple Data Stream); (iii) MIMDS (Multiple Instruction Multiple Data Stream). Machines may also be compared according to the rate at which instructions are carried out. Also they may be compared according to the power of the individual instructions in their instruction set, whereby "power" refers to the complexity of the operation which is initiated by a single instruction. Finally, they may be compared according to the ease with which programs may be coded by the user. Usually such comparisons lead to unsatisfactory and/or confusing results. The particular user who is considering acquiring or fabricating a machine wishes to know how effective this machine will be in performing specific tasks. Whether it is a SIMDS of MIMDS machine, whether it carries out a picture point operation (pixop) in 100 ns or 50 ns, or whether it has a repretoire of 20 or 100 instructions is relatively unimportant. What is important is usually the number of hours of programming required to program for a particular task and the number of hours of execution time required once this task has been programmed and is being utilized. For this reason, the user develops benchmark tasks with which to test various machines. He attempts to design these benchmarks to be representative of the major types of operations which he will use during the life of the machine.

It is the purpose of this chapter to propose a benchmarking methodology which would be applicable to the special-purpose parallel processors which are now evolving. Both the architecture and language for several of these processors were described at the workshops on High-Level Languages for Image Processing and New Computer Architectures and Image Processing upon which this book is based. Unfortunately, in the time available for the preparation of this chapter it was impossible to develop benchmarks and test all the machines described. Rather, three machines were selected, plus another machine built during the summer of 1980 at Carnegie-Mellon University and the University of Pittsburgh, using a single benchmark test. This chapter, therefore, reports upon the initiation rather than the completion of a comparison between these parallel processing machines. It is hoped that the methodology developed and demonstrated in this chapter will inspire (1) the completion of a more comprehensive set of benchmark tasks and (2) the rigorous testing of machines already developed and also those currently on the drawing board. In the sections below a short description is given for each of the machines selected, of the characteristics of the benchmark tasks, and of the preliminary results obtained. The reader should keep in mind that the results reported are the outcome of the early part of the author's investigation and may still contain some inaccuracies.

Table 1

Language	Machine	Institution
MORPHAL	AT4	Centre for Mathematical Morphology (France)
CAP4	CLIP4	University College London (England)
C3PL	CYTOCOMPUTER	Environmental Research Institute of Michigan (USA)
DAP FORTRAN	DAP	ICL (England)
GLOL	DIFF 3	Coulter Biomedical Research Corp. (USA)
—	DIP	Delft University of Technology (Netherlands)
—	FLIP	Institute for Information Processing (Germany)
INTRAC	GOP	Linkoeping University (Sweden)
—	IP	Hitachi Central Research Laboratory (Japan)
PPL	PICAP II	Linkoeping University (Sweden)
—	PPP	Toshiba Research and Development Centre (Japan)
—	SYMPATI	CERFIA–UPS (France)

2. MACHINES TESTED

Four machines were selected for investigation: two which are commercial machines; two, research devices. All machines are SIMDS and were selected for their capability in executing instructions peculiar to the benchmark task chosen. The commercial machines are the diff3 of Coulter Electronics Inc. (Hialeha, Florida, USA) and the DAP (Distributed Array Processor) of International Computers Ltd. (London, UK). The two research devices are the PPP (Parallel Pattern Processor) of the Toshiba Research and Development Centre (Kawasaki city, Kanagawa, Japan) and the PHP (Preston-Herron Processor) of Carnegie-Mellon University and the University of Pittsburgh. The performance of these machines is shown in Fig. 1 in terms of pixops per second as a function of the year in which they first came on-line. These machines are of interest in that their performance establishes a clear trend line, starting with the earlier Perkin-Elmer machines, namely, CELLSCAN and GLOPR, which shows a seven order-of-magnitude improvement in performance over the past 20 years (a factor of 60 every five years). The MPP (Massively Parallel Processor) now being developed by Goodyear, Akron, under contract to the United States National Aeronautics and Space Administration (USNASA) which is scheduled to be operational in 1982 also falls on this trend line. Unfortunately, our request to USNASA for information on the capabilities of the MPP in executing the benchmark task described below were unanswered. Therefore, this chapter provides no information on the performance of that system.

Performance, of course, can be quantitated using other measures than pixops per second. Another measure frequently used relates to cost-effectiveness, namely, pixops per second per dollar. CELLSCAN, built in

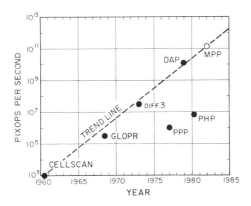

Fig. 1. Picture point operation (pixop) rates for various parallel image processing machines.

1961 by the Navigation Computer Corporation for the Perkin-Elmer Corporation, cost $20,000. Using 1980 dollars, this amounts to approximately 2.10^{-2} pixops per second per dollar. In contrast, the DAP at a cost of approximately 600,000 British pounds, prices out at about 10,000 pixops per second per dollar. The corresponding numbers for GLOPR, PPP, PHP, and diff3 are 4:100:1000:1000 respectively (and for MPP; 30,000). Although these figures are of interest for comparison purposes they do not tell the whole story.

2.1 The diff3

The diff3 contains the GTP (Golay Transform Processor, an updated version of GLOPR) which is a parallel processor for binary images. The host computer is a Data General Nova-2. The system contains four dedicated image memories (each 64×64 with six bits per picture element). Any one of these image memories may be accessed by a high-speed thresholding circuit to deliver a binary image to one of the four additional 64×64 binary image memories. These binary memories are 8-rank shift registers whose contents is circulated at 25 ns to the GTP which extracts each picture point and its 8-neighbourhood in order to carry out cellular logic by table lookup. This operational speed is achieved by using 8 processing elements (PEs) running in parallel with look-up tables read out in 200 ns.

The diff3 was initially developed by the Perkin-Elmer Corporation for the purpose of carrying out automated microscopy in hematology laboratories where it is now used to process and identify approximately 5500 white blood cell images per hour. This system is now manufactured and sold by the Coulter Biomedical Research Corporation (Concord, Massachusetts, USA), a division of Coulter Electronics Inc. Although, at present, the diff3 is specialized for automated hematology, the GTP is generally programmable in GLOL (GLOPR Operating Language) for a multiplicity of image processing tasks. The system is more completely described by Graham and Norgren [1].

2.2 The DAP

The DAP is a 64×64 array of PEs which may be thought of as 4096 64×64 bit planes. The basic cycle time for the array is 200 ns, resulting in a pixop time of 50 ps. The DAP is the fastest full-array processor now in existence. It is programmable in a special language called DAPFORTRAN and can perform certain simple operations (AND, OR, EXOR) in a single cycle time. The programming language is suitable for the development of image processing algorithms either involving Boolean or numerical processing. Also, an assembly language permits the construction of macros which could be assembled in a specialized programming language tailored to a particular application area (much like GLOL). In this case the user program would consist of a number of assembly language subroutines, each being mainly a sequence of macro calls, plus a number of DAPFORTRAN subroutines.

The DAP has been in commercial manufacture by International Computers Ltd since the end of the 1970s. It is completely described in Chapter 22 of this book by David J. Hunt.

2.3 The PPP

The PPP is the image processing unit for the TOSPICS (Toshiba Pattern Information Cognitive System). The PPP is a microprogrammable cellular logic processor which has several basic image processing functions reduced to simple ultra-high-speed hardware. These modules conduct (1) two-dimensional convolution, (2) point mapping, (3) linear coordinate transformation, (4) logical filtering, (5) histogram generation, (6) region labelling, and (7) pixel operations (gradient, etc.).

All image memories are 512×512 with neighbourhood functions carried out over sub-arrays which may be 3×3 up to 8×8 in size. The general design philosophy is that each module conduct operations with a pixop time of 1 μs. The configuration of the PPP as an instruction set processor [2] makes it unique in the group of four machines investigated in the process of preparing this chapter. Further details on the design of the PPP are provided by Mori *et al.* [3].

2.4 The PHP

The PHP is a low-cost (approximately $5,000) logical transform processor designed to be used as a peripheral to the Perkin-Elmer 3200-series mini-computers. It was constructed in late 1980 in a joint effort between Carnegie-Mellon University, the University of Pittsburgh and the Perkin-Elmer Corporation. It resides on a standard ULI (Universal Logic Interface) Perkin-Elmer printed circuit board plugged into the main-frame of the host. As with the GTP of diff3, it handles binary images only. Unlike diff3, whose image memory is specifically limited to a line length of 64 pixels, the PHP is capable of processing images whose dimensions $N \times M$ may be varied from 16 to 4096 (modulo 16). Image length and width variations are electronically pro-grammable. The PHP accepts image data at 16 megapixels per second. It has 16 PEs which operate in parallel using 16 identical look-up tables. Before conducting each logical image transform the lookup tables are electronically loaded from the host computer. Thus PHP is electronically reconfigurable each image processing cycle. The results image is passed back to the host computer at the DMA rate, resulting in an equivalent pixop time of 125 ns.

3. BENCHMARK TASK

Studies at Carnegie-Mellon University have shown that a multi-threshold logical transform program for object detection, counting, and sizing is useful in the analysis of microscopical images of human tissue (see Fig. 2). In addition

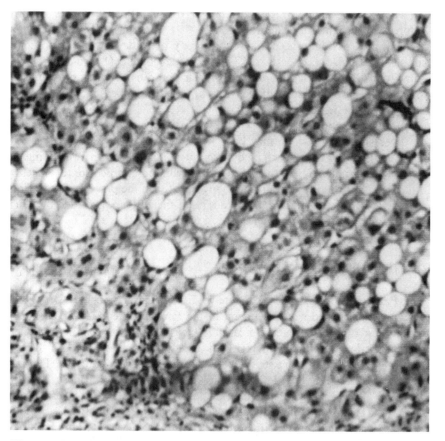

Fig. 2. Microscope image of 0.5 mm × 0.5 mm human liver tissue section (10 μm thick) digitized using green illumination by the Automatic Light Microscope Scanner model 2 at the Jet Propulsion Laboratory in a 512 × 512 array at 8 bits per picture element.

it has been determined that tissue architecture, i.e. the arrangement of cells within the tissue matrix, may be quantitated by histogramming the arcs of the exoskeleton of the nuclei of the cells which make up the tissue itself. Because of our familiarity with the details of this image-processing operation, because of its general utility in biomedical image analysis, and because it incorporates a wide variety of basic image-processing operations, it was selected as the benchmark task. The performance of this task was then categorized into four separate sub-tasks.

3.1 Image I/O

The first step in any image processing operation is to transfer the image itself from an input device (scanner, magnetic tape, disc, etc.) to the core of either the host computer or to dedicated memory in the parallel processing machine. In some cases, of course, this operation includes analog to digital conversion.

3.2 Nuclei extraction

For purposes of this benchmark a specific routine was selected for finding the cell nuclei and reconstructing them so as to preserve size information. First the range of the pixel value histogram is calculated and divided into 16 equal intervals. These intervals are then used to threshold the initial image in order to produce 16 binary images (see Fig. 3). Eight of these binary images, starting with that corresponding to a threshold at the highest optical density, are processed for the purpose of extracting cell nuclei. To carry out this processing the binary image is reviewed and all objects with a maximum Feret's diameter of 10 picture elements are extracted. Results of this processing on all 8 binary images are then ORed. An illustration is provided in Fig. 4.

3.3 Nuclei histogram

The next step in the benchmark task is to carry out a histogram of the object (cell nuclei) in the image which results from ORing. The number of bins in the histogram is set to 4.

3.4 Exoskeleton analysis

Using the complement of the image resulting from the operation carried out in section 3.2, the exoskeleton is generated. A histogram of the arcs of the

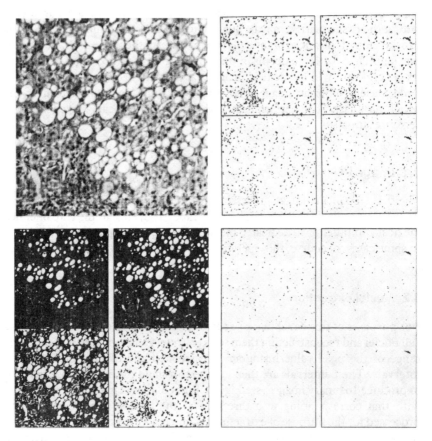

Fig. 3. The image shown in Fig. 2 (upper left corner) may be thresholded to produce a multiplicity of binary images (12 are shown above).

exoskeleton is then produced again using 4 bins. This operation is illustrated in Fig. 5.

4. RESULTS

For each of the parallel processing machines investigated, a program was generated for conducting the benchmark task, the program was executed, and timing data was gathered. In one case (DAP) timing data was estimated. The results for the four machines is presented in Table 2 and for DAP, in Table 3. Appendices A, B and C provide the computer programs. The languages used

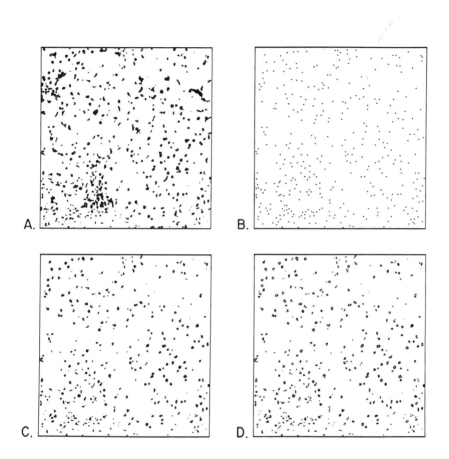

Fig. 4. Nuclei extraction from a binary image produced by thresholding: (A) input binary image, (B) residues of objects having a maximum Feret's diameter less than or equal to 10 picture elements, (C) replication from residues using the input image as template, (D) objects in the fourth histogram bin EXORed with their residues.

are GLOL for the diff3 and SUPRPIC for the PHP. These languages are described in the literature [4, 5]. The program for PPP is given in the unnamed language for this machine whose command repertoire is given in the paper by Mori *et al.* [3]. For both diff3 and the PPP individual execution times are given for all commands.

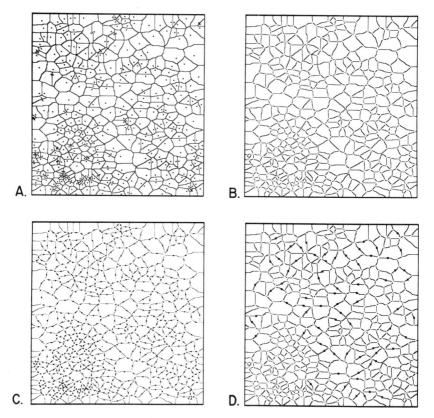

Fig. 5. Exoskeleton analysis: (A) nuclei exoskeleton, (B) nodes removed, (C) residues of the arcs, (D) arcs in the fourth histogram bin ORed with their residues.

5. CONCLUSIONS

Although the benchmark described is far from representative of all types of image processing, this investigation has led to certain obviously helpful results and, perhaps, will lead to the development of a more rigorous set of benchmark tests and their application to parallel processors for image analysis. As is indicated in the following tables the rate of execution of a benchmark task does not follow proportionately the pixop rate. For example, DAP which executes 20 thousand pixops per microsecond is only 10 times faster than diff3 in performing the benchmark task although diff3 conducts only 40 pixops per microsecond. Clearly this is due to the fact that the diff3 is optimized for this type of benchmark task while the DAP is not. Continuing this comparison,

Table 2
Table 2
Estimated execution times for four machines

Coulter diff3 (Nova-2/GTP programmed in GLOL)

Sub-task	64 × 64 (ms)	4096 × 4096 (s)	Percentage
Image I/O	3.0	12.3	4.2
Nuclei extraction	42.5	147.1	59.2
Nuclei histogram	2.5	10.2	3.5
Exoskeleton analysis	23.8	97.5	33.1
Total	71.8	294.1	100.0

International Computers Ltd (DAP programmed in DAP-FORTRAN)

Sub-task	64 × 64 (ms)	4096 × 4096 (s)	Percentage
Image I/O	(1.02)	(4.18)	(14.9)
Nuclei extraction	2.10	8.60	30.2
Nuclei histogram	0.42	1.72	6.1
Exoskeleton analysis	3.40	13.93	50.0
Total	6.94	28.43	100.0

Toshiba TOSPICS (TOSBAC-40C/PPP using command repetoire)

Sub-task	512 × 512 (s)	4096 × 4096 (s)	Percentage
Image I/O	0.04	2.24	—
Nuclei extraction	12.44	796.16	58.5
Nuclei histogram	—	—	—
Exoskeleton analysis	8.78	561.92	41.4
Total	21.26	1360.32	100.0

Carnegie-Mellon/University of Pittsburgh (PE8/32/PHP in SUPRPIC)

Sub-task	512 × 512 (s)	4096 × 4096 (s)	Percentage
Image I/O	80.0	5120.0	22.5
Nuclei extraction	160.0	10240.0	45.0
Nuclei histogram	16.0	1024.0	4.5
Exoskeleton analysis	100.0	6400.0	28.0
Total	356.0	22784.0	100.0

Table 3
DAP estimated execution time (64 × 64 image)

Operation		μs
Histogram (32 levels)		500
Smooth and calculate thresholds		—
Nuclei extraction loop (do for 8 thresholds)		
Threshold	7	
De-noise	10	
Reduce 4 steps	156	
Identify residues	3	
Expand residues } AND (6 steps) }	24	
OR with next	1	
	201 × 8 =	1608
Size histogram loop (do for 4 bins)		
Reduce 1 step	39	
Identify residues	3	
Count residues	60	
EXOR	1	
	103 × 4 =	412
Exoskeleton analysis		
NOT	1	
Generate exoskeleton	2496	
Remove nodes	20	
Size histogram loop (do for 4 bins)		
Reduce 4 steps	156	
Identify residues	3	
Count residues	60	
EXOR	1	
	220× 4 =	880
	3397 × 1 =	3397
		5917
TOTAL		

note that, although the PHP can conduct 8 pixops per microsecond in comparison with one pixop per microsecond for the PPP, the PPP is in fact 200 times *faster* than the PHP in conducting the benchmark task. This is due to two reasons. The existing PHP software driver at present does not take full advantage of the real-time features of the Perkin-Elmer operating system. Because of this software limitation, the PE 8/32 cannot keep up with the PHP. More importantly, however, the PPP, being an instruction set processor, with the labelling routine as one of its hardware instructions, overwhelms the PHP in the execution of this subtask. In fact (see Fig. 6), the PHP is only slightly faster than GLOPR in this particular operating environment.

Finally, let us make a few systems-oriented remarks concerning the benchmark task itself. The reason that all tables show the execution time for a 4096 × 4096 image is that this is the size of the tissue sample (in picture elements) which is usually examined by the pathologist in the course of his analysis. Typically, the pathologist will spend 5 min on this task. However, it is likely that the pathologist will spend less than one hour a day observing tissue samples so that only one or two thousand samples are observed per annum. From Fig. 6 we see that neither GLOPR or PHP could handle the pathologist's work load. The PPP could readily handle this work load if it could perform continuously. However, it requires almost half an hour to do one specimen. This would not satisfy the demands of the pathology laboratory in urgent cases where only 5 mins are available for a rapid decision on pathology during surgery. On the other hand the diff3 is nicely matched to this task at exactly 5 min per specimen. The DAP is so fast that it could conduct the analysis of one specimen 10 times faster than the pathologist and could handle the work load of an entire health centre (several hospitals) per year. This speed is

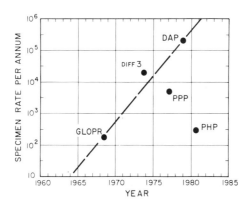

Fig. 6. Rates for performing the benchmark task for complete liver tissue specimens (4.0 mm × 4.0 mm) for various parallel image processing machines.

unlikely to be required and the cost penalty (the DAP is 10 times more expensive than the diff3), in this case, would not justify the cost of a DAP installation economically. However, if the DAP could also handle other image processing tasks for the health centre (such as computed tomography reconstruction, radiological image analysis, etc.), it might prove economical.

This chapter, therefore, has shown that there are many considerations beyond the simple architecture and pixop rate of an image processing machine which must be carefully studied before procuring or building such a system. It is important for designers of these systems to carefully analyse in advance a multiplicity of image processing tasks in terms of the demands of the user community. In doing so and in testing various design concepts by benchmarking, it should be possible to match future machines to their environment in a wiser and more flexible manner.

ACKNOWLEDGEMENTS

The author would like to acknowledge the support of the National Research Council (UK) and the Consiglio Nazionale delle Ricerche (Italy) as well as the efforts of Mr D. J. Hunt (International Computers Ltd., UK), Drs M. Kidode (Toshiba Research and Development Centre, Japan), D. Graham (Coulter Biomedical Research Corp., USA) and J. Herron (University of Pittsburgh, USA) in performing the work reported. The assistance of Mr J. Farley, a senior in the Biomedical Engineering Program at Carnegie-Mellon University, who performed benchmark tests is also appreciated. Further acknowledged is the work of Ms Rosalie Ballentine (Executive Suite, Tucson, USA) in typing the major portion of the manuscript as well as the help of Ms Beverly Douglas (Carnegie-Mellon University) and Ms Kathy Sidorovich (University of Pittsburgh). Mr G. Arnold and his staff, Mr G. Thomas, Mr M. Westfall and Ms Mary Adams did the line drawings and photographic illustrations.

REFERENCES

[1] Graham, M. D. and Norgren, P. E. (1980). "The diff3 Analyzer: A parallel/serial Golay image processor". In *Real-time medical image processing*, Onoe, M., Preston Jr., K. and Rosenfeld, A. (eds). Plenum Press, New York.
[2] Reddy, D. R. and Hon, R. W. (1980). "Computer architecture for vision". In *Computer vision and sensor-based robots*. Plenum Press, New York.
[3] Mori, K-i., Kidode, M., Shinoda, H. and Asada, H. (1979). "Design of local parallel pattern processor for image processing". Proc. AFIPS Conf. **47**, 1025–1031.

4] Preston Jr., K. (1972). "Use of the Golay logic processor in pattern recognition studies using hexagonal neighborhood logic". In *Computers and automata*, pp. 609–624. Polytechnic Press, New York.

5] Preston Jr., K. (1980). "Image manipulative languages: a preliminary survey". In *Pattern recognition in practice*, Kanal, L. N. and Gelsema, E. S. (eds). North-Holland, Amsterdam.

APPENDIX A

GLOL Benchmark

8THRS: CLEAN: GET RESIDUES: REPLICATE: ACCUMULATE: SIZE CELLS:
: EXOSKELETON: SIZE ARCS

Execution time (ms)		GLOL command	Explanation of command
—	—	IA 0,32,P	*Define 32-element array and init. to zero
—	—	IA 0,16,Q,R	*Define two 16-element arrays and init.
3.0	3.0	HI P	*Histogram image in array P
—	—	SM P,P,10,10,T	*Smooth P 10 cycles with threshold 10
—	—	AR Q(16)=P(T)−P(0)	*Store range of histogram in Q(16)
—	—	AR Q(15)=Q(16)/15	*Store threshold increment in Q(15)
—	—	DO I/2,8	*Calculate 8 equispaced
—	—	AR J=I−1	*Threshold levels and
—	—	AR Q(I)=Q(J)+Q(15)	*Store in Q(1) . . . Q(8)
—	—	END LOOP	
0.1	3.1	C=0	*Clear logical image array C
—	—	DO I/1,8	*Initiate nuclei extraction
3.0	—	AC A,Q(I)	*Threshold image and store in A
0.2	—	A=[G(A)A]3−6,2	*De-noise contents of A
0.1	—	B=A	*Copy A into B
1.2	—	B=[G'(B)B]1−4,4,3	*Partial reduction to residues
0.1	—	B=[G(B)B]0,,	*Save residues only in B
0.6	—	B=[B+G(B)A]1−13,I	*Replicate nuclei from residues
0.1	—	C=C+B	*Accumulate nuclei in C
—	45.5	END LOOP	
0.1	45.6	D=0	*Clear logical image array D
—	—	DO I/9,12	*Start sizing nuclei
0.3	—	C=[G'(C)C]1−4,1,3	*Partial reduction to residues
0.1	—	B=[G(C)C]0,,	*Store residues of C in B
—	—	CO X	*Count
—	—	AR Q(I)=Q(I)+X	*Store count in Q(I)
0.1	—	D=D+B	*Accumulate residues in D
0.1	—	C=C−B	*Remove counted residues from C
—	48.0	END LOOP	
0.1	48.1	D=D'	*Complement field of residues
9.6	57.7	D=[G'(D)D]1−4,I,3	*Generate exoskeleton
0.1	57.8	D=[G(D)D]11−12,,	*Remove nodes from exoskeleton
1.2	59.0	D=[G'(D)D]1−4,4,3	*Generate residues from short arcs
		DO I/1,4	*Start exoskeleton histogram

Appendix A *Continued*

Execution time (ms)		GLOL command	Explanation of command
0.1	——	A=[G(D)D]0,,	*Copy residues of D in A
—	——	COUNT X	*Count
—	——	AR R(I)=R(I)+X	*Store count in R(I)
0.1	——	D=D−A	*Remove counted residues from D
3.0	——	D=[G′(D)D]1−4,10,3	*Generate new residues in D
—	71.8	END LOOP	
—	——	OU 1,Q(9)−Q(12)	*Output to device 1 the nuclear size
—	——	OU 1,R(1)−R(4)	*and Exoskeleton ARC histograms

APPENDIX B

PPP Benchmark (512 × 512)

8 THRS: LABEL: ACCUMULATE: SIZE CELLS: EXOSKELETON: SIZE ARCS:

Execution time (ms)		PPP command	Explanation of command
—	—	INTEGER WND(4)/1,1,512,512/	*Defines processing window
—	—	INTEGER HSTBUF(256), DCVTBL(256)	*Dimension arrays
—	—	INTEGER THD(8)	
35	35	CALL TVIN (M1,WND)	*Input to M1 form TV scanner
262	297	CALL HIST(M1,WND,HSTBUF)	*Compute image histogram
—	—	MAX=0	*Generate threshold values in THD
—	—	MIN=256	
—	—	DO 5 I=1,256	
—	—	IF(HSTBUF(I) .GT .MAC) MAC=HSTBUF(I)	
—	—	IF(HSTBUF(I) .LT .MIN) MIN=HSTBUR(I)	
—	—	5 CONTINUE	
—	—	DEL=(MAX−MIN)/16	
—	—	DO 6 I=1,8	
—	—	6 THD(I)=MIN+DEL*(I−1)	
262	559	CALL IMCLR(M4,0)	*Set M4 to zero
—	—	DO 10 I=1,8	*Labeling of threshold image

Appendix B *Continued*

Execution time (ms)		PPP command	Explanation of command
262	—	CALL LABEL (M2,M1,M3, WMD,8,GE,THD(I))	
262	—	CALL HIST (M2, WND,HSTBUF)	*Compute size histogram
—	—	(code for generating DCVTBL)	
—	—		
262	—	CALL DCV(M3,M2,WND,DCVTBL)	*Extract objects of certain size
393	—	CALL OR(M4,M3,M4,WND)	*OR results in M4
—	12087 10	CONTINUE	
		(code for generating the size histogram)	
393	12480	CALL NOT (M4,WND)	*Invert M4
8384	20864	CALL THIN (M3,M4,WND,32)	*Generate exoskeleton
		(code for removing nodes from the exoskeleton)	
		(code for histogramming arc lengths)	

APPENDIX C

SUPRPIC6 Benchmark

8THRS: CLEAN: GET RESIDUES: REPLICATE: ACCUMULATE: SIZE CELLS
: EXOSKELETON: SIZE ARCS

0101000809101100	*THRESHOLD AT LEVELS 08–11
0409011001000000	*COPY LU09 BUF01 INTO LU10 BUF01
0910010202090001	*REMOVE NOISE
0910011605040601	*GET RESIDUES OF CELL NUCLEI
0410011002000000	*COPY RESULTS INTO LU10 BUF02
0910020100090001	*REMOVE RESIDUES IN LU10 BUF02
0810011002100200	*RESIDUES ONLY IN LU10 BUF02
0510021002000000	*INVERT LU10 BUF02
0910020107090001	*AUGMENT RESIDUES
0510021002000000	*INVERT LU01 BUF02
0609011002100200	*AND WITH ORIGINAL
0510021002000000	*INVERT LU10 BUF02
0910020405040501	*AUGMENT IN SUBFIELDS
0510021002000000	*INVERT LU10 BUF02
0609011002100200	*AND WITH ORIGINAL

Appendix C *Continued*

```
0510021002000000   *INVERT LU10 BUF02
0910020405040501   *AUGMENT IN SUBFIELDS
0510021002000000   *INVERT LU10 BUF02
0609011002100200   *AND WITH ORIGINAL
0510021002000000   *INVERT LU10 BUF02
0910020405040501   *AUGMENT IN SUBFIELDS
0510021002000000   *INVERT LU10 BUF02
0609011002100200   *AND WITH ORIGINAL
0510021002000000   *INVERT LU10 BUF02
0910020405040501   *AUGMENT IN SUBFIELDS
0510021002000000   *INVERT LU10 BUF02
0609011002100200   *AND WITH ORIGINAL
0510021002000000   *INVERT LU10 BUF02
0910020405040501   *AUGMENT IN SUBFIELDS
0510021002000000   *INVERT LU10 BUF02
0609011002100200   *AND WITH ORIGINAL
0510021002000000   *INVERT LU10 BUF02
0910020405040501   *AUGMENT IN SUBFIELDS
0510021002000000   *INVERT LU10 BUF02
0609011002100200   *AND WITH ORIGINAL
0510021002000000   *INVERT LU10 BUF02
0910020405040501   *AUGMENT IN SUBFIELDS
0510021002000000   *INVERT LU10 BUF02
0609011002100200   *AND WITH ORIGINAL
0510021002000000   *INVERT LU10 BUF02
0910020405040501   *AUGMENT IN SUBFIELDS
0510021002000000   *INVERT LU10 BUF02
0609011002100200   *AND WITH ORIGINAL
0410021003000000   *ACCUMULATE RESULTS IN LU10 BUF03
```

This code is repeated for each threshold.

```
0910030405040600   *PARTIAL REDUCTION TO RESIDUES
0410031004000000   *COPY RESULTS INTO LU10 BUF04
0910040100090000   *REMOVE RESIDUES IN LU10 BUF04
0810031004100400   *RESIDUES ONLY IN LU10 BUF04
1800100400000000   *COUNT RESIDUES IN 1004
0910030405040600   *PARTIAL REDUCTION TO RESIDUES
0410031004000000   *COPY RESULTS INTO LU10 BUF04
0910040100090000   *REMOVE RESIDUES IN LU10 BUF04
0810031004100400   *RESIDUES ONLY IN LU10 BUF04
1800100400000000   *COUNT RESIDUES IN 1004
0910030405040600   *PARTIAL REDUCTION TO RESIDUES
0410031004000000   *COPY RESULTS INTO LU10 BUF04
```

Appendix C *Continued*

0910040100090000	*REMOVE RESIDUES IN LU10 BUF04
0810031004100400	*RESIDUES ONLY IN LU10 BUF04
1800100400000000	*COUNT RESIDUES IN 1004
0910030405040600	*PARTIAL REDUCTION TO RESIDUES
0410031004000000	*COPY RESULTS INTO LU10 BUF04
0910040100090000	*REMOVE RESIDUES IN LU10 BUF04
0810031004100400	*RESIDUES ONLY IN LU10 BUF04
1800100400000000	*COUNT RESIDUES IN 1004
0510031003000000	*INVERT FINAL RESIDUES
0910030107090001	*AUGMENT RESIDUES
0910036405040501	*GENERATE PARTIAL EXOSKELETON
0510031003000000	*INVERT EXOSKELETON
0910030100090000	*REMOVE NOISE
0510031003000000	*INVERT
0910036405040501	*COMPLETE EXOSKELETON
0410031004000000	*COPY INTO LU10 BUF04
0910040108060001	*MARK HIGH ORDER NODES
0510041004000000	*INVERT
0910040107090001	*AUGMENT NODES
0610031004100400	*REMOVE NODES
0910041208040600	*REDUCE EXOSKELETON ARCS TO RESIDUES
0410041001000000	*COPY INTO LU10 BUF01
0910010100090000	*REMOVE RESIDUES IN LU10 BUF01
0810031001100100	*RESIDUES ONLY IN LU10 BUF01
1800100100000000	*COUNT RESIDUES
0910041208040600	*REDUCE EXOSKELETON ARCS TO RESIDUES
0410041001000000	*COPY INTO LU10 BUF01
0810031001100100	*RESIDUES ONLY IN LU10 BUF01
1800100100000000	*COUNT RESIDUES
0910041208040600	*REDUCE EXOSKELETON ARCS TO RESIDUES
0410041001000000	*COPY INTO LU10 BUF01
0910010100090000	*REMOVE RESIDUES IN LU10 BUF01
0810021001100100	*RESIDUES ONLY IN LU10 BUF01
1800100100000000	*COUNT RESIDUES
0910046408040600	*REDUCE EXOSKELETION ARCS TO RESIDUES
0410041001000000	*COPY INTO LU10 BUF01
0910010100090000	*REMOVE RESIDUES IN LU10 BUF01
0810031001100100	*RESIDUES ONLY IN LU10 BUF01
1800100100000000	*COUNT RESIDUES
2000000000000000	*END PROCEDURE

Subject Index